In Honor of Prof. Liritzis Ioannis: Essays in Archaeology & Archaeometry and the Hellenic Contribution to Egyptology

In Honor of Prof. Liritzis Ioannis: Essays in Archaeology & Archaeometry and the Hellenic Contribution to Egyptology

Editors

Grigorios Tsokas
Nikolaos Lazaridis
Omar Abdel-Kareem

MDPI • Basel • Beijing • Wuhan • Barcelona • Belgrade • Manchester • Tokyo • Cluj • Tianjin

Editors

Grigorios Tsokas
Aristotle University of Thessaloniki
Greece

Nikolaos Lazaridis
California State University
USA

Omar Abdel-Kareem
Cairo University
Giza

Editorial Office
MDPI
St. Alban-Anlage 66
4052 Basel, Switzerland

This is a reprint of articles from the Special Issue published online in the open access journal *Heritage* (ISSN 2571-9408) (available at: http://www.mdpi.com).

For citation purposes, cite each article independently as indicated on the article page online and as indicated below:

LastName, A.A.; LastName, B.B.; LastName, C.C. Article Title. *Journal Name* **Year**, *Volume Number*, Page Range.

ISBN 978-3-0365-3399-5 (Hbk)
ISBN 978-3-0365-3400-8 (PDF)

Contents

About the Editors

Grigorios Tsokas, Professor of Applied Geophysics Research, Department of Geology, Laboratory of Exploration Geophysics, School of Geology, Aristotle University of Thessaloniki, Greece. Research interests and current work include the development of archaeological prospection data acquisition; Mediterranean geoarchaeology; processing and interpretation techniques; potential fields theory and practice; and complex attributes analysis in archaeological exploration geophysics. Member of the European Academy of Science and Arts.

Nikolaos Lazaridis is Professor of Ancient Mediterranean History at California State University, Sacramento. He is an expert in comparative ancient literature, ancient epigraphy, and Egyptian culture. Member and chief epigrapher of the North Kharga Oasis Survey team. Current projects include "The monograph Storytellers at work: Crafting characters in ancient Egyptian stories"; "Epigraphy in North Kharga Oasis's Darb Ain Amur"; and "Aegyptiaca in the Greek world".

Omar Abdel Kareem is Professor of Conservation of (mainly) Egyptian Antiquities, Conservation Dept., Faculty of Archaeology, Cairo University, Cairo, Egypt. His research interests include the analysis and conservation of archaeological and culture heritage objects; conservation sciences; the restoration of material culture; textiles; organic; inorganic materials; and cultural management.

heritage

MDPI

Editorial

Essays in Archaeology and Archaeometry and the Hellenic Contribution to Egyptology

Nikolaos Lazaridis [1,*], Omar Abdel-Kareem [2] and Grigorios Tsokas [3]

1 Department of History, College of Arts & Letters, California State University, Sacramento, CA 95819, USA
2 Conservation, Conservation Department, Faculty of Archaeology, Cairo University, Cairo 12613, Egypt; omaa67@yahoo.com
3 Laboratory of Exploration Geophysics, Department of Geology, School of Geology, Aristotle University of Thessaloniki, GR-541 24 Thessaloniki, Greece; gtsokas@geo.auth.gr
* Correspondence: lazaridi@csus.edu

Abstract: The contemporary trend of research projects and works are presented on selective issues of archaeometry, archaeology and Egyptology. The current status in research in the area of SE Mediterranean on cultural heritage and archaeological/historical reflections alone and/or coupled with archaeological sciences of eleven papers are placed within an updated frame. The results concern a variety of selected topics critically presented. The topics touch on the cultural astronomy, the ancient textiles and masonries and the physico-chemical and biological investigations, the socio-political issues of Egyptian Ramesside era, revisiting the inscription of an Egyptian statuette, and the valuable information extracted from rock graffiti in north Kharga, Egypt.

Keywords: archaeoastronomy; masonry; conservation; embroidery; Coptic; Ramesside; dyes; graffiti; mummy

Citation: Lazaridis, N.; Abdel-Kareem, O.; Tsokas, G. Essays in Archaeology and Archaeometry and the Hellenic Contribution to Egyptology. *Heritage* **2022**, *5*, 402–408. https://doi.org/10.3390/heritage5010023

Received: 18 January 2022
Accepted: 15 February 2022
Published: 22 February 2022

The interdisciplinary field of Heritage culture, archaeology, history of art, coupled with natural sciences, is undoubtedly a breakthrough in the advancement of cultural heritage studies and sustainability [1,2].

The regional development of heritage studies has begun to attract a great deal of research interest, and SE Mediterranean studies with a focus on Egypt are rich in novel results. Research from these disciplines converge, granting us a higher level of knowledge of heritage from prehistory to historical times. Aspects of tangible and intangible cultural heritage concerning interconnected ancient societies and their activities, development and ingenuity in the SE Mediterranean regions, the cradle of civilizations, are presented in the present volume.

This Special Issue contains eleven articles touching on a variety of selected topics from these subjects; in particular, two on the archaeoastronomy for the earliest mechanical device to the temples at Bekaa, Lebanon; five on the study of dyes in Coptic Textiles and the physico-chemical and biological investigations of ancient textiles and stone masonries and mummies of historic and archaeological significance for conservation purpose; then, a detailed description and interpretation of rare Coptic embroidery traced in the history of art heritage objects; the socio-political issues of the way Egyptians perceived the boundaries of their land and the impact of Egyptian colonialism during the Ramesside period; a reassessment of the inscription of the Egyptian statuette in the Heraklion Archaeological Museum for a more plausible translation; and last, the valuable information extracted from rock graffiti in north Kharga, Egypt.

The Special Issue is made to honor Professor Ioannis Liritzis' for his retirement from the University of the Aegean, Greece—he is now a distinguished professor at Henan University, China—for his long and prolific academic career which has advanced research in archaeology, cultural heritage and the archaeological sciences, involving multiple initiatives in the archaeology of Egypt (ushering the first chair in Egyptology in University of the Aegean), and the wider Eastern Mediterranean archaeology and cultural heritage (www.liritzis.eu, accessed on 2 November 2021; https://archlab.aegean.gr/, accessed on 25 January 2022; https://en.wikipedia.org/wiki/Egyptology, accessed on 10 February 2022). Prof. Liritzis' research output refers to a wide spectrum of specialized subjects which are tackled by most of this volume's articles.

Since the early days of mankind, humans have looked to the stars for answers to their biggest questions. The skyscape has been the source of inspiration, imagination, for science and philosophy. The determination of time and identification of human or animal images with them has exerted a strong influence on the human cultures throughout the millennia. The four solar stands, the lunar phases, the planetary orbits, the rise and setting of bright stars and the constellations have triggered past societies to determine the time for daily works. Worldwide examples from the five continents on these aspects are outlined, focusing on the determination of time with solar stands in daylight, the lunar periodic movements, and bright stars and associated configurations (constellations). In essence, the determination of time is deemed necessary for many daily aspects, such as agriculture (sowing, harvesting, cultivation), maritime voyages, rituals and religious festivities, the preservation of the four (climatic) seasons, art, and related myths. In particular, the natural and/or constructed devices in the determination of time and geographical location (sundials, hydraulic and portable devices, topographical markers) including intangible oral documentation of cultural astronomy of aboriginals provides clues on the intangible heritage and other social aspects of the old world [3].

Archaeoastronomy has advanced since the times of the simple and somewhat naïve studies or uncontextualized alignments and can be considered today as a complete, interdisciplinary science tightly connected to cognitive archaeology [4].

Analysis of specific orientations of temples, buildings, altars, natural markers is made to connect them to specific deities, sacred attributes, rites, delivering oracles, major cultural events [4–14] and for the determination of time [15–18].

Concerning the trans-scientific and interdisciplinary approach of conservation science to valuable heritage objects (organic and inorganic origin) and monuments, the current research has produced effective strategies. These follow certain sequential steps; from the characterization of weathering, corrosion and microbial inhibition to development of compatible materials, with emphasis on nanotechnology, to prevent further deterioration and for consolidation to digital recording for museum exhibit. These involves spectroscopic techniques and experimental simulation of aging and conservation tests to define proper materials for conservation. The most significant and most recent contributions have been made to heritage materials of the SE Mediterranean cultures [19–38].

On the history and archaeology side, the tapestry was known as a mirror of civilization, because many tapestries represented scenarios of everyday life. Tapestry may be similar to painting with yarn, depicting an endless picture. The finest articles are seen in Coptic Egypt around the fifth and the sixth centuries A.D. [39]. It became an important issue in the 9th century and later the Iconoclastic controversy in the Near and Middle East and the West, especially during the late Carolingian and Ottonian periods, in a great number of iconographic variations. In some instances, it might be accompanied by the Virgin, St. John, the thieves, or other figures associated with the Passion, recorded on various materials (ivory, wood, textile, stone). Thus, rare objects of that period are of importance in the heritage field of art history and these need proper investigation [39–43].

The socio-political issue of land boundaries was and still is a thorny topic. Egyptians' perception of the boundaries of their land and the Egyptian and colonial encounters has been discussed in the turmoiled region of near East [44,45].

Remoteness in time is one of the main difficulties we encounter when we look for sources of information about the Old Kingdom. Thus, any decipherment of Egyptian hieroglyphic inscriptions are valuable sources of information of past Egyptian dynasties concerning rituals, religion, war affairs, afterlife, etc. [46,47].

The valuable information extracted from rock graffiti in Egypt is gaining worldwide attention.

Sites such as Al-Qarta and Abu Tanqoura, north of Komombo town, Nasr Salam, Hula Rock 1 and 2 are ways of communication as texts, images, or signposts, or rock art; they are adding their voices to the muted interactions of all desert travelers who chose to carve messages on such rocks. Whatever these and others in the world mean, these are artistic expressions which bear a message; therefore, it is worthwhile to explore and preserve them [48–54].

In recent years, new research has revived the interest in the unsolved problems posed by the Baalbek monuments, Bekaa Valley, Lebanon. Such problems include dating construction phases, determining relationships with the landscape, and discerning the nature of the cult practiced. In the paper by Magli [55] archeoastronomy is used to suggest that the plan of the Temple of Jupiter was a unified one, conceived under Herod the Great. Magli also argues that the cult was strongly connected to the renewal of seasonal cycles and proves the existence of an orientation custom which appears to originate in Baalbek and to inform all these sacred places.

The earliest mechanical device in the world, dated 2150 years ago, the Antikythera Mechanism found in a shipwreck in the Aegean Sea early 20th century is the oldest extant complex geared device, an amazing analogue computer. The device was operated manually by a user, setting a date in a dial. The intended calculations were made using a set of at least 39 gears while the results were displayed on several scientific scales. This Greek device was used to calculate astronomical phenomena, such as solar and lunar eclipses, constituting confirmation of the written sources regarding the Pan-Hellenic games and the accuracy of its predictions is demonstrable. Efstathiou et al. [56] have developed an application which simulates the exact operation of the physical model of the mechanism. From all the tests that have been performed, it is concluded that the mechanism accurately predicts the astronomical phenomena of the future and confirms with astonishing accuracy the astronomical phenomena of the past.

The role played by the clay minerals and microorganisms in the deterioration process of architectural units is a hot topic in the conservation and restoration of ancient buildings. Building materials (mainly mortars and plasters) from the church of Virgin Mary, Wadi El-Natrun region, Egypt, were examined using spectroscopic techniques (X-ray diffraction, X-ray fluorescence and scanning electron microscope with energy dispersive spectroscopy, by Moussa and Roshdy [57]. It was found that wall gypsum mortars in the church contain halloysite as a dominant clay mineral while plaster is clay free; concerning microorganisms, the fungal flora Aspergillus glaucus represent the most dominant fungi.

The assessment, modeling, and strengthening of historic masonry walls of multi-leaf systems essentially require suitable knowledge of their construction technology, typology, geometrical characteristics, and the properties of their components. In this context, a comprehensive structural survey of multiple-leaf walls of medieval historic buildings in Cairo was performed by Amer et al. [58]. Three construction hypotheses of multiple-leaf stone masonry walls are presented considering weak, thick, and monolithic core infill layers. This novel work provided a proper characterization as a prerequisite for determining the most suitable materials and techniques for further strengthening interventions.

Ismael et al. [59] report their research results on the comparative inhibition of three materials of nano zinc oxide (ZnO-NPs), Ceratophyllum demersum and 4-chloro-m-cresol for the microcide on the screaming Mummy in Egyptian Museum store mummies. This screaming mummy suffers from stains due to microbial infection. Microorganisms, isolated from the degraded mummy, were identified with an optical microscope and ribosomal ribonucleic acid (rRNA) analysis to guarantee identification accuracy. Results indicated that the bacteria in the mummy are *Bacillus jeotgali, Kocuria turfanensis, Microbacterium imperial, Micrococcus luteus* and *Bacillus megaterium*. Fungi are *Monascus pallens* and *Rhizopus oryzae*. The results of minimum inhibitory concentration (MIC) illustrated that the best concentrations for the bio treatment of isolated microorganisms is plant extract (*Ceratophyllum demersum*) followed by 4-chloro-m-cresol and finally nano zinc oxide.

The research team led by Amer et al. [60] presented further development of multiple-leaf stone masonry walls in Egyptian heritage monuments to correctly define the pathology and determine the appropriate interventions for its conservation and preservation. Thus, comprehensive studies on its building materials were carried out implemented to enrich documentation, conservation and restoration issues of this type of wall. The fundamental physical and mechanical properties of the masonry elements were examined by incorporating stone blocks, mortars and core-infill materials utilizing a large range of complementary investigation and analysis techniques, including polarizing microscopy, X-ray diffraction (XRD), thermal analysis (TG/DTA), and environmental scanning electron microscope (ESEM) attached to an EDX unit. The outer leaves of the majority of the multiple-leaf stone-masonry walls in medieval architectural heritage were mainly built of well-dressed limestone blocks with nearly uniform dimensions, while the inner-core layer was usually built of stone-rubble infill with bending lime-based mortar. The uniaxial compressive strengths of core infill and lime-based mortar of the embedded joints are shown to be 85 and 92.5% lower than the limestone units of the outer layer, respectively. Moreover, experimental observations indicate that the inner core layer exhibits the highest porosity values; consequently, deteriorated, loose and cohesionless core infill could greatly affect the durability and thermal resistivity of this kind of wall. The results provide scientific support for investigating the overall structural behaviour of this type of walls and for decision-making in future conservation and restoration strategies.

Karapanagiotis et al. [61] presented new evidence on the identification of Dyes from twelve ancient Egyptian textiles (4th–5th c. AD) from the Museum of Faculty of Archaeology, Cairo University. Applying high performance liquid chromatography coupled to a diode-array-detector (HPLC-DAD), they investigated samples which were extracted from ancient Egyptian textiles (4th–5th c. AD. They discovered madder, weld (*Reseda luteola* L.) and indigoid dyes which can be either indigo (Indigofera species and other) or woad (*Isatis tinctoria* L.). These identifications were achieved, as alizarin, purpurin, rubiadin (madder components), apigenin, luteolin, chrysoeriol (weld components), indigotin and indirubin (indigo/woad components) were detected in the chromatograms of the investigated samples. Additionally, ellagic acid was detected, implying the presence of tannin sources. In some of the studied samples, madder was the only detected dye. It is reported that these samples were treated with a madder source which was rich in purpurin and poor in alizarin. Madder dyes rich in alizarin (probably originated from *R. tinctorum*) were detected only in samples treated with madder–tannin mixtures. Finally, it is noted that the potential use of other madder species (*Galium* and *Asperula species*) by the Egyptian dyers cannot be ruled out by the chemical results.

State-of-the-art surface-profiling equipment has been used by Gleeson [62] on the Egyptian statuette, AM Heraklion no. 95, and has established previously unknown hieroglyphic signs, so that the dedicator's second title can now be read, with some confidence, as *Maker of the Wadjet Standard*.

Lila de Chaves [63] presents the "Crucifixion with the twelve Apostles", a unique Coptic embroidered panel, was on display at the Benaki Museum (Athens, Greece), representing of the "Crucifixion" with Christ in the center and six Apostles on either side, standing next to each other in frontal poses, is quite a rare one. This unique representation of the Crucifixion with the twelve Apostles, which also involves the Ascension, is a one-of-a-kind compositional formula representing Christ's death as a triumph over Death, emphasizing, along with the other factors, its non-Chalcedonic origin. A detailed description and interpretation of the inscription, written in at least three languages. Stylistically, in many details, one can trace Armenian connotations; in addition, there is a Syrian influence especially in the rendering of the Apostles' faces. The inscription constitutes a synthesis of letters from different alphabets. The particular personality of Coptic art prevails, with its abstract, conventional forms and its pronounced mysticism and symbolism, that, nonetheless, remains deeply Christian. As for its place of origin, our working hypothesis would assign it to the Low Egypt region of Fayoum dated at the 13th–14th century.

The work of Xekalaki [64] aims to define the way Egyptians perceived the boundaries of their land and reassesses the impact of Egyptian colonialism during the Ramesside period (c. 1292–1069 BCE). During this era, expansive wars, diplomatic action and land administration/governance reforms led Egypt to control a large part of modern Israel, Palestine, Lebanon and Syria. Seeking to explore whether the use of modern terms on ancient Egypt may be an anachronism, this paper reviews the scholarship on (a) Egyptian records documenting conquests and (b) contextual archaeological evidence from the southern Near East itself. This review highlights differences between modern and ancient conceptions of land domination. Finally, Egyptian border-related terms are used in a strictly local symbolic cultural context but not in the one of international diplomacy. As for Egypt's boundary, it was mostly formed as a buffer zone rather than a borderline.

In the course of the last ten years, the North Kharga Oasis–Darb Ain Amur Survey team, led by Salima Ikram (American University in Cairo), has been exploring a network of interconnected desert paths in Egypt's Western Desert, known as Darb Ain Amur. These marked paths run between Kharga Oasis and Dakhla Oasis, linking them to Darb el-Arbain, a notorious caravan route facilitating contacts between Egypt and sub-Saharan Africa since prehistoric times. Lazaridis [65] reports on two rock graffiti in North Kharga, Egypt, focusing on the types of information one may extract from such ancient epigraphic materials, viewing these as a sustained flow of information between carved messages that probably created a safety net that eased desert travelers' discomfort by making them feel as important contributors to the ongoing efforts to make Western Desert's terrain more hospitable and more navigable.

All the above articles highlight significant novel archaeological and archaeological sciences issues in the SE Mediterranean and Egypt in particular.

Author Contributions: Conceptualization, N.L. and O.A.-K.; methodology, N.L.; O.A.-K., G.T.; software, N.L.; O.A.-K., G.T.; validation N.L.; O.A.-K., G.T.; formal analysis, N.L.; investigation, N.L.; O.A.-K., G.T.; resources N.L.; O.A.-K., G.T.; data curation, N.L.; O.A.-K., G.T.; writing—original draft preparation, N.L.; writing—review and editing, N.L.; O.A.-K., G.T.; visualization, N.L.; O.A.-K., G.T.; supervision, N.L, and O.A.-K.; project administration, N.L.; All authors have read and agreed to the published version of the manuscript.

Funding: This research received no external funding.

Institutional Review Board Statement: Not applicable.

Informed Consent Statement: Not applicable.

Data Availability Statement: Not applicable.

Conflicts of Interest: The authors declare no conflict of interest.

References

1. Liritzis, I.; Laskaris, N.; Vafiadou, A.; Karapanagiotis, I.; Volonakis, P.; Papageorgopoulou, C.; Bratitsi, M. Archaeometry: An overview. *Sci. Cult.* **2020**, *6*, 49–98.
2. Liritzis, I.; Korka, E. Archaeometry's Role in Cultural Heritage Sustainability and Development. *Sustainability* **2019**, *11*, 1972. [CrossRef]
3. Liritzis, I.; Vlachos, A. *Skyscape Impact to Cultural Astronomy*; Liritzis, I., Ed.; Interdisciplinary Essays in Archaeology, Cultural Heritage and Environment, Book Series: Interdisciplinary Contributions to Archaeology; Springer-Nature: Dordrecht, The Netherlands, 2022; in press.
4. Magli, G. Archaeoastronomy. In *Introduction to the Science of Stars and Stones*; Springer: New York, NY, USA, 2015.
5. Sweatman, M.B.; Tsikritsis, D. Decoding Göbekli tepe with archaeoastronomy: What does the fox say? *Mediterr. Archaeol. Archaeom.* **2017**, *17*, 233–250.
6. Lluis i Ginovart, I.; Ugalde-Blázquez, I.; Lluis-Teruel, C. Gisemundus and the orientation of the romanesque churches in the spanish pyrenees (11th–13th centuries). *Mediterr. Archaeol. Archaeom.* **2021**, *21*, 205–214.
7. Castro, B.; Liritzis, I.; Nyquist, A. Oracular Functioning and architecture of five ancient Apollo Temples through archaeastronomy: Novel approach and interpretation, Nexus Network Journal. *Archit. Math.* **2015**, *18*, 373–395.

8. Liritzis, I.; Bousoulegka, E.; Nyquist, A.; Castro, B.; Alotaibi, F.M.; Drivaliari, A. New evidence from archaeoastronomy on Apollo oracles and Apollo-Asclepius related cult. *J. Cult. Herit.* **2017**, *26*, 129–143. [CrossRef]
9. Liritzis, I.; Castro, B. Delphi and Cosmovision: Apollo's absence at the land of the hyperboreans and the time for consulting the oracle. *J. Astron. Hist. Herit.* **2013**, *16*, 184–206.
10. Steiner, G.F. Archaeoastronomy and Bedouin star-lore in the rock art of the Negev desert. *Mediterr. Archaeol. Archaeom.* **2017**, *17*, 243–260.
11. Liritzis, I.; Vassiliou, H. Archaeoastronomical orientation of seven significant ancient Hellenic temples. *Archaeoastron. Int.* **2003**, *17*, 94–100.
12. Munro, A.M.; Gullberg, S.R. Skyscape archaeology as a road to cultural insight. *Mediterr. Archaeol. Archaeom.* **2018**, *18*, 491–497.
13. Zhang, L. Tomb orientation and posthumous visit to the capital of the supernatural world: The Guo cemetery at Sanmenxia, Henan, China. *Mediterr. Archaeol. Archaeom.* **2020**, *20*, 145–161.
14. Liritzis, I.; Vassiliou, H. Does sunrise day correlate with eastern orientation of Byzantine Churches during significant solar dates and Saint's day name? A preliminary study. *Byz. Zeitscrift* **2006**, *99*, 523–534.
15. Pliakos, A.A. Are there 12 or 10 the minoan solar 'months'? A reassessment. *Mediterr. Archaeol. Archaeom.* **2021**, *21*, 227–236.
16. Pérez-Valcárcel, J.; Pérez Palmero, V. Orientation of the churches in the hispanic medieval castles. *Mediterr. Archaeol. Archaeom.* **2021**, *21*, 241–257.
17. Efstathiou, K.; Basiakoulis, A.; Kokkinos, N. Determining the celebration of the next pythian games using the antikythera mechanism, given that they continued to be celebrated non-stop to this day. *Sci. Cult.* 2022, *in press*.
18. Voulgaris, A.; Mouratidis, C.; Vossinakis, A.; Bokovos, G. Renumbering of the antikythera mechanism saros cells, resulting from the saros spiral mechanical apokatastasis. *Mediterr. Archaeol. Archaeom.* **2021**, *21*, 107–128.
19. Abo-Taleb, T. Documenting oil paintings by fingerprint brushstroke application to antique painting in the Egyptian museum of modern art. *Egypt. J. Archaeol. Restor. Stud.* **2019**, *9*, 141–154. [CrossRef]
20. Michalopoulou, A.; Maravelaki, P.; Stefanis, N.A.; Theoulakis, P.; Andreou, S.; Kilikoglou, V.; Karatasios, I. Evaluation of nanolime dispersions for the protection of archaeological clay-based building materials. *Mediterr. Archaeol. Archaeom.* **2020**, *20*, 221–242.
21. Alves, C.; Figueiredo, C.A.M.; Sanjurjo-Sánchez, J.; Hernández, A.C. Salt Weathering of Natural Stone: A Review of Comparative Laboratory Studies. *Heritage* **2021**, *4*, 1554–1565. [CrossRef]
22. Acri, G.; Testagrossa, B.; Faenza, P.; Caridi, F. Spectroscopic analysis of pigments of the antonello gagini annunciation's sculptural marble group, church of St. Theodore martyr (Bagaladi, reggio Calabria, Italy): Case study. *Mediterr. Archaeol. Archaeom.* **2020**, *20*, 1–5.
23. Capitelli, F.; Dida, B.; Ventura, G.D.; Baldassarre, F.; Capelli, D.; Senesi, G.S.; Mele, A.; Siliqi, D. Functional Nano-Hydroxyapatite for Applications in Conservation of Stony Monuments of Cultural Heritage. *Proceedings* **2020**, *62*, 11. [CrossRef]
24. Chyla, J.M. How Can Remote Sensing Help in Detecting the Threats to Archaeological Sites in Upper Egypt? *Geosciences* **2017**, *7*, 97. [CrossRef]
25. Hemeda, S.; Osman, T. Impacts of fire on historic stone masonry structures: Physicochemical analysis and application to the al-musafirkhana palace (Cairo). *Mediterr. Archaeol. Archaeom.* **2021**, *21*, 191–209.
26. Theologitis, A.; Kapridaki, C.; Kallithrakas-Kontos, N.; Maravelaki-Kalaitzaki, P.; Fotiou, A. Mortar and plaster analysis as a directive to the design of compatible restoration materials in frangokastello (Crete). *Mediterr. Archaeol. Archaeom.* **2021**, *21*, 109–120.
27. Amer, O.; Abdel-Aty, Y.; Abd El-Hady, M.; Aita, D.; Torky, A.; Hussein, Y.M. Multiscientific-based approach to diagnosis and characterization of historic stone-masonry walls: The mausoleum of al-imam al-shafi'i, Cairo (Egypt). *Mediterr. Archaeol. Archaeom.* **2020**, *20*, 67–82.
28. Arlt, T.; Mahnke, H.E.; Siopi, T.; Menei, E.; Aibéo, C.; Pausewein, R.R.; Reiche, I.; Manke, I.; Lepper, V. Absorption edge sensitive radiography and tomography of Egyptian Papyri. *J. Cult. Herit.* **2019**, *39*, 13–20. [CrossRef]
29. Abdallah, M.; Abdrabou, A.; Kamal, H.M. Multiscientific analytical approach of polychrome Greco-Roman palette applied on a wooden model naos: Case study. *Mediterr. Archaeol. Archaeom.* **2020**, *20*, 45–65.
30. Ali, M.F.; El-Shafey, H.; Marey Mahmoud, H. Multianalytical techniques of al-bīmāristān al-mu'ayyidi mural painting at historic Cairo: Contribution to conservation and restoration. *Sci. Cult.* **2021**, *7*, 33–47.
31. Al-Gaoudi, H.A. Painted ancient Egyptian mummy cloth of khonsuemrenep from bab el-gasus excavation: Scientific analysis and conservation strategy. *Sci. Cult.* **2020**, *6*, 49–64.
32. Abdelrahim, S.A.; Elnagar, K. Analytical study and conservation processes of late period limestone canopic jar: Case study. *Sci. Cult.* **2020**, *6*, 25–38.
33. Zanini, A.; Trafeli, V.; Bartoli, V. The laser as a tool for the cleaning of Cultural Heritage. *IOP Conf. Ser. Mater. Sci. Eng.* **2008**, *364*, 012078. [CrossRef]
34. Mohamed, M.I.; Wael, S.M.; Hamdy, M.M. Experimental study for evaluation of paraloid® B72 and its nanocomposite with nano TiO_2 and nano ZnO for consolidation of pottery samples. *Sci. Cult.* **2021**, *7*, 101–111.
35. Borsoi, G.; Santos Silva, A.; Menezes, P.; Candeias, A.; Mirão, J. Analytical characterization of ancient mortars from the archaeological roman site of Pisões (Beja, Portugal). *Constr. Build. Mater.* **2019**, *204*, 597–608. [CrossRef]
36. Abdelmoniem, A.M.; Mahmoud, N.; Samaha, S.H.; Mohamed, W.S. Characterization of the best consolidation material for black resin for the late period coffin. *Sci. Cult.* **2020**, *6*, 1–7.

37. Jimenez-Lopez, C.; Jroundi, F.; Pascolini, C.; Rodriguez-Navarro, C.; Piñar-Larrubia, G.; Rodriguez-Gallego, M.; González-Muñoz, M.T. Consolidation of quarry calcarenite by calcium carbonate precipitation induced by bacteria activated among the microbiota inhabiting the stone. *Int. Biodeterior. Biodegrad.* **2008**, *62*, 352–363. [CrossRef]
38. Afifi, H.A.M.; Rushdya Rabee, A.H.; Menofy, S.M. An experimental study for consolidation of archaeological cartonnage using klucel g and chitosan, with nanocalcium hydroxide. *Sci. Cult.* **2021**, *7*, 49–68.
39. Shailaja, D.; Naik, J.; Wilson, A. *Surface Designing of Textile Fabrics*; New Age International Pvt Ltd. Publishers: New Delhi, India, 2006.
40. Meinardus, O. The collection of Coptica in the QASR of the Monastery of St. Antony. Bull. *Société Archéologie Copte* **1980**, *18*, 272.
41. Schiller, G. *Iconography of Christian Art*; New York Graphic Society: New York, NY, USA, 1972; Volume 2, pp. 323–324.
42. Noble, T.F.X. *Images, Iconoclasm, and the Carolingians*; University of Pennsylvania Press: Philadelphia, PA, USA; Routledge: Oxfordshire, UK, 2021.
43. Karydis, C.; Oikonomou, A.; Konstanta, A. The unpublished Coptic textiles of the monastery of St. John the theologian: Preliminary results of previous alterations and scientific analysis. *Mediterr. Archaeol. Archaeom.* **2019**, *19*, 133–142.
44. Koch, I. *Colonial Encounters in Southwest Canaan during the Late Bronze Age and the Early Iron Age; Culture and History of the Ancient Near, East*; Frahm, E., Randall, W., Baruch, C., Theo, H., van den Hout, P.J., Winteret, I.J., Eds.; Koninklijke Brill NV: Leiden, The Netherlands, 2021; Volume 119, pp. 1–11.
45. Moreno García, J.C. Recent Developments in the Social and Economic History of Ancient Egypt. *J. Anc. Near East. Hist.* **2014**, *2*, 231–261. [CrossRef]
46. Adkins, L.; Adkins, R. *The Keys of Egypt: The Obsession to Decipher Egyptian Hieroglyphs*; Harper Collin: New York, NY, USA, 2000.
47. Tara, S. Writing the Past: Ancient Egypt through the Lens of Medieval Islamic Thought. In *Arabic Humanities, Islamic Thought: Essays in Honor of Everett K. Rowson*; Lowry Joseph, E., Toorawa Shawkat, M., Eds.; Brill Academic Pub: Brill Leiden, The Netherlands, 2017.
48. El-Badry, A.E.H.A.; Ali, M.F.; Ismail, B.M. Assessment of the aging treatments of sandstone greywacke rock art (Wadi Hammamat) by petrography, Sem, Xrd, Edx. *Sci. Cult.* **2019**, *5*, 37–48.
49. Sternberg, J. Were Indians responsible for all of the peterborough petroglyphs? *Sci. Cult.* **2019**, *5*, 61–66.
50. Gingrich, A. Rock art from west and south-west Arabia: Socio-cultural anthropology's insights for the region's eastern transition zones. *Mediterr. Archaeol. Archaeom.* **2017**, *17*, 61–73.
51. Bednarik, R.G.; Khan, M. The Rock Art of Southern Arabia reconsidered. *Adumatu J.* **2009**, *20*, 7–20.
52. Miqdadi, R.H. ABJAD numerals as an absolute dating method: Forts from al-ain, UAE. *Mediterr. Archaeol. Archaeom.* **2020**, *20*, 273–289.
53. Mustafa, M.H. Human or centipede? Egyptian influence on the Nabataean culture. *Mediterr. Archaeol. Archaeom.* **2019**, *19*, 39–50.
54. Rzepka, S. *Who, Where and Why: The Rock Graffiti of Members of the Deir El-Medina Community*; University of Warsaw, Inst of Archaeology, Egypt & Nubia: Warsaw, Poland, 2014.
55. Magli, G. Archaeoastronomy of the Temples of the Bekaa Valley. *Heritage* **2021**, *4*, 1526–1537. [CrossRef]
56. Efstathiou, K.; Efstathiou, M.; Basiakoulis, A.; Kokkinos, N. The Antikythera Mechanism: The Prove of the Accuracy of the Astronomical Calculations Based on It. *Heritage* **2021**, *4*, 3848–3878. [CrossRef]
57. Moussa, A.; Roshdy, M. Monitoring Coptic Masonry Affected by Clay Minerals and Microorganisms at the Church of Virgin Mary, Wadi El-Natrun (Egypt). *Heritage* **2021**, *4*, 4056–4067. [CrossRef]
58. Amer, O.; Aita, D.; Mohamed, E.K.; Torky, A.; Shawky, A. Multi-Leaf Stone Masonry Walls in Egypt: A Legend. *Heritage* **2021**, *4*, 2763–2791. [CrossRef]
59. Ismael, S.; Omar, A.; Maher, M. Comparative Inhibition Study by Nanomaterial, Plant Extract and Chemical Microcide on the Screaming Mummy in Egyptian Museum Store. *Heritage* **2021**, *4*, 2481–2493. [CrossRef]
60. Amer, O.; Aita, D.; Mohamed, E.K.; Torky, A.; Shawky, A. Experimental Investigations and Microstructural Characterization of Construction Materials of Historic Multi-Leaf Stone-Masonry Walls. *Heritage* **2021**, *4*, 2390–2415. [CrossRef]
61. Karapanagiotis, I.; Abdel-Kareem, O.; Kamaterou, P.; Mantzouris, D. Identification of Dyes in Coptic Textiles from the Museum of Faculty of Archaeology, Cairo University. *Heritage* **2021**, *4*, 3147–3156. [CrossRef]
62. Gleeson, L. A Short Note on the Knossos Statuette Inscription. *Heritage* **2021**, *4*, 3186–3192. [CrossRef]
63. Lila de Chaves, L. An Investigation on a Coptic Embroidered Panel from the 13th Century "Crucifixion with the Twelve Apostles" (Benaki Museum, Athens). *Heritage* **2021**, *4*, 4335–4343. [CrossRef]
64. Xekalaki, G. On Borders and Expansion: Egyptian Imperialism in the Levant during the Ramesside Period. *Heritage* **2021**, *4*, 3938–3948. [CrossRef]
65. Lazaridis, N. "Like Wringing Water from a Stone!" Information Extraction from Two Rock Graffiti in North Kharga, Egypt. *Heritage* **2021**, *4*, 2253–2260. [CrossRef]

heritage

Article

An Investigation on a Coptic Embroidered Panel from the 13th Century "Crucifixion with the Twelve Apostles" (Benaki Museum, Athens)

Lila de Chaves [1,2]

1 Coptic Textiles Collection of the Benaki Museum, 1 Koumbari Str., 106 74 Athens, Greece; lildechav@yahoo.fr
2 Member CIETA, Centre Internationale D'etudes de Textiles Anciens, 34 Rue de la Charite, 69002 Lyon, France

Abstract: The "Crucifixion with the twelve Apostles", a unique Coptic embroidered panel, was on display at the Benaki Museum (Athens, Greece). The representation of the "Crucifixion" with Christ in the center and six Apostles on either side, standing next to each other in frontal poses, is quite a rare one. This rare iconographic image of the twelve Apostles could be linked to the Ascension or the Pentecost. This unique representation of the Crucifixion with the twelve Apostles, which also involves the Ascension, is a one-of-a-kind compositional formula representing Christ's Death as a triumph over Death, emphasizing, along with the other factors, its non-Chalcedonic origin. Moreover, the interpretation of an inscription, written in at least three languages embroidered in black silk thread, is a matter which confuses the issue even more. In the present study, we will attempt a comprehensive investigation, a detailed description, and interpretation of this rare iconography, based on written and iconographic evidence traced in the history of art heritage objects.

Keywords: crucifixion; apostles; inscription; silk; embroidery; Greek; Coptic; monastery

Citation: de Chaves, L. An Investigation on a Coptic Embroidered Panel from the 13th Century "Crucifixion with the Twelve Apostles" (Benaki Museum, Athens). *Heritage* **2021**, *4*, 4335–4343. https://doi.org/10.3390/heritage4040239

Academic Editors: Christofilis Maggidis, Nikolaos Lazaridis and Omar Abdel-Kareem

Received: 30 August 2021
Accepted: 8 November 2021
Published: 13 November 2021

1. Introduction

This rare heritage collection piece, *The Crucifixion with the 12 Apostles*, a Coptic embroidered panel including a multilingual inscription, has been exposed in the Benaki Museum, Athens, until its refurbishment in 1989 (Figure 1).

Figure 1. "The Crucifixion with the twelve Apostles". A rare and mysterious embroidered silk panel (283 cm × 48 cm) of the late Coptic monastic period 13th–14th cent., Benaki Museum, Athens, Greece.

The "Crucifixion with the twelve Apostles" (Figure 1), a unique embroidered panel (283 cm × 48 cm) of late Coptic art, was purchased in May 1929 by Anthony Benaki, the founder of the Museum, from Nahman, an antique dealer, for the price of 100 sterling pounds. It appears listed for the first time in the 1935 Catalogue of the Benaki Museum, with the simple description "The Crucifixion with the Twelve Apostles" (inv. ΓΕ7148). The description and dating initially was made by Otto Meinardus [1], who describes it as an embroidery of "the Crucifixion on silk with six Apostles on each side, one of whom slightly elevated", as well as by the director of the Museum, late Prof. A. Delivorrias [2], with the only addition "Tentatively assigned to the 7th century".

The theological interpretation of the death of Christ, since the early Christian period, has been articulated through the Crucifixion in works of minor art (manuscript illustration, weaving, metalwork, engraving), such as the 5th century ivory relief from North Italy, the Crucifixion on the Monza ampulla (British Museum) (Figure 2). For the evolution of the iconography of the Crucifixion see [3,4].

Figure 2. The Crucifixion on a 5th century ivory relief from North Italy, British Museum; Judas hanged, M for Mary, J for St. John and S for soldier ([4] and 16; based on https://www.britishmuseum.org/collection/object/H_1856-0623-5, BM No. 1856,0623.5).

It became an issue in the 9th century and later the Iconoclastic Controversy in the East, Syria, Armenia, Egypt, and the West, especially during the late Carolingian and Ottonian periods, in a great number of iconographic variations; in some instances, it might be accompanied by the Virgin, St. John, the thieves, or other figures associated with the Passion. The representation of the "Crucifixion" in question, with Christ in the center and six Apostles on either side, standing next to each other in frontal poses, is quite a rare one. On a Greek seal stone, 3rd c. AD, in the British Museum, the earliest similar type is found, yet this type, as attested by Gertrud Schiller, most likely comes from heretical circles [4] (Figure 321). A similar arrangement of the Apostles can be found in a scene of the Ascension, a fresco in the Baouit/Bawit Monastery V II [5] (p. 42), a deserted Monastery (known also as Monastery of St. Apollo, Deir Abu Abullu), lying in the Egyptian desert, facing the fertile plain, about fifteen kilometers from Dairut, north of Aysut (Assiut).

In the present research, note this unique heritage object, the Coptic embroidered panel, is revisited, and the intriguing issue of the position and artistic expression of the Apostles reassessed [1].

2. Technical Context

Initially, the fibers of the textile and those of the embroidered figures have been examined by microtomic section at the ITF (Institut Textile de France) Paris laboratories. The results were as follows: *Textile*: purplish brown and orange fibers, silk. *Embroidery*: Crucifix, Apostles: undyed white fibers, cotton. Inscription, crosses: black-dyed fibers, cotton (particularly good and well-preserved product).

2.1. *Technique of the Textile*

The crucifixion is a rare complete example of late Coptic monastic embroidery, despite the damage to parts of the textile (263 cm × 48 cm). The narrative scene is embroidered on a purplish-brown silk field in plain weave, which is decorated with three woven bands, perpendicular to the embroidery, (263 cm × 7.5 cm) of unspun orange silk threads in Samit

(2Z1). Under the iconographic scene on which an inscription is embroidered lengthwise, the third orange Samit band is cut in the middle. This verifies that the textile was much wider. An inlay-brocade pattern with the same orange unspun silk threads decorates the larger upper band of the embroidered panel. It comprises a repetition of geometric designs, such as the diaper scheme and a motif reminiscent of the candelabra tree ending to a smaller "head" upper element. This is an important motif of the Umayyad period, which was quickly adopted by the Coptic weavers of the time, and has persisted throughout the period of Mamluks [6] (pp. 10, 12, 53, 55, 145). (For more extensive information on Textiles, see [7] (Figure 8, p. 35–49)). Coincidentally, the salvage at the top of the embroidered scene (at the right side of the textile, on which the embroidery was made), has been preserved and shows, along with the weaving, the close fitted dents of a proper reed. This leads us to presume that the textile must have not been weaved before the late 12th–13th centuries, taking into consideration the technical evolution of the draw loom. Hence, we are dealing with a silk textile, which is expertly braided, but, nevertheless, associated with a draw loom mainstream production, which delimits us to date it between the 13th–14th centuries AD.

2.2. *The Technique of Embroidery*

A portion of the material is stitched with the scene that occupies the space between the orange bands in the self-couched and knotted stem stiches, which were not only characteristic of the late Coptic period but also used commonly in Medieval Eastern and Western Monasteries [5] (p. 646). The embroidered version of the "Crucifixion with the twelve Apostles" shows a primitive and inferior technique. Despite that, the effort of the embroiderer is obvious, which is the portrayal of various iconographical details. This can be seen on the garments of the Apostles, the details of which have been rendered only by means of using a different direction of the stitch. This is a consequence of the fact that the embroidery was mainly embroidered with white threads, except for some details which were rendered in red, brown, blue, salmon pink, and black (using cotton dyed threads). Some of these colors have faded with time. The textile at the places of the embroidered figures has shrunk and unsightly folds appear, which were created by the reaction of the cotton threads, possibly due to the fact that the embroidery remained in a humid environment for a long time. The oxidation has caused some decay in the wears and holes on part of the textile.

3. Stylistic Description and Discussion

The architectural layout of a Coptic Monastery church is formed by three chapels. The "Haikal" is the central one, forming essentially the sanctuary which contains the high altar and is usually closed by an embroidery curtain like the one we are examining. Usually, in the various Christian iconographic depictions of the crucifixion, the twelve Apostles are never presented all together since Judas had already died. If this occurs, in a rare case, then Matthew is selected as the 12th Apostle, in the place of Judas. From Chapter B of the Acts of the Apostles, it is derived that the depiction with the twelve Apostles is iconographically established with the Pentecost (see, Figure 1 and Figure 5 below). That is, when the number of the twelve Apostles is complete, and all of them together receive *The Holy Spirit*, they start preaching in the different languages. This is also mentioned in the *First Epistle* (6, 4) of Apostle Paul to the Corinthians.

The Crucifixion with the twelve Apostles (Figure 1), although rendered stylistically in a primitive and non-naturalistic way, with a total absence of any decorative elements, has, undoubtedly, a monumental nature. The Christ, nailed on the cross, occupies the center of the panel, dividing the scene in two symmetrical parts in an unusual compositional configuration. On either side stand six Apostles, one next to each other, in frontal poses, all of equal size except for the one far right, who is smaller. Only the Crucifix is larger; thus, the scene acquires importance and becomes centrobaric. Neither the Apostles nor the figure of Christ seem to be supported firmly on any pictorial ground. They all look rather suspended in the narrow space of the textile, in an endless arrangement; thus, it is in accordance with the aesthetic expression of the two-dimensional surface, on which everything is suspended in a world that is transcendental and symbolic, so dear to the Copts. Christ has his eyes closed in death. In the East, the iconography of the dead Christ first appeared after the iconoclast controversy in 843 AD, as it did in the West (School of Reims) [see Radbertus in Paulus [8], (p. 9)].

A depiction which was first traced on an icon from Mount Sinai of the 8th century as presented by Weitzmann and Chatzidakis is nailed to a T-shape cross (*Crux Comissa patibulata*), or cross of St. Antony of Egypt (Figure 3). It is also used as the head of Pastoral staffs for the actual or symbolic support of ecclesiastics in the Orthodox and Coptic church [9] (p. 18). The cross is embroidered in purplish brown with some details in black, which evidently, although rendered in a naive way, play the role of the chiaroscuro around Christ's figure. His arms are outstretched and nailed; blood is also traced in salmon pink [10] (Figures 24–27) [11]. During the early Byzantine period, in the first representations of the Crucifixion on minor art objects, Christ appeared nailed to the cross with four nails: two for his hands and two for each of his legs, which lied frontally one next to the other nailed separately on the cross, like on a gold crucifixion "encolpion". (See Dumbarton Oaks from early byzantine 6th to 7th centuries. Gold 8 cm × 5.30 cm, BZ1937.24.).

Figure 3. The crucifix stands in the middle of the scene, Christ's eyes closed in death, being nailed on a T-shaped cross (*Crux comissa patibulata*) with only three nails and wearing a loincloth (perizoma).

Both his legs, the right one on top of the left, are nailed to the upright of the cross with only one nail and are not supported on a suppedaneum. This type of Crucifixion with only three nails has its roots in the North, and it first appeared in the 12th century. This representation shows a new outlook on the Passion, Arma Christi [4] (p. 146) frequently appearing in many regions. Eventually, it travelled to the East with the Crusades, during the 12th–13th centuries [3] (p. 412).

Christ has a dark beard embroidered in black and lighter hair embroidered in brown. According to G. A. Sotiriou, these iconographic features are of a Jewish type [12] (p. 109). His body is portrayed frontally, rendered in a primitive style, with obvious disproportion, his legs being much shorter than his torso. His posture is heavy and stiff without any realistic leaning either to the right (Byzantine type) or to the left, *contraposto* (western type).

He is wearing a perizoma (loincloth), which replaced the *colobium* or sleeveless Chiton (Tunic) from the 9th century onwards (Figure 3).

An interesting scene with the Christ wearing a colobium at a later time [4] (p. 100) is found on a fresco from the Episcopi church in Eurytania (central Greece), middle 12th century in the Byzantine Museum in Athens. In either of the two lateral compositions flanking the Crucifix, the Apostles are shown as huge images. They all have bell-shaped (Figure 4) conventionally rendered bodies, a style which is properly Coptic. Such evidence is found in various iconographic sources, on works of minor arts, and on icons and frescoes, as have been presented in the articles and works of Maggy Rassart-Debergh, Piotr O. Scholz, P. P. du Bourguet, M.-H. Rutschowskaya, and L. A. Hunt, to mention only a few of the many experts. Some of the Apostles are wearing colobia and other chitons with a small cross on the breast and *himatia* (BIKOYKAION in Coptic, that is, the Byzantine Phelonion or the Catholic Chasuble). (For an extensive description of Coptic ecclesiastical vestments, see [13].

Figure 4. One of the three of the bell-shaped Apostles having their hands raised in a gesture of testimony. Each one is holding an orb with an inscribed red cross in their left hand.

It is noticed that their heads are larger in proportion to the rest of the bodies. Their eyes are rendered with just a black line; the same goes for the eyebrows and the nose, which is rendered in an oblong V shape. Their mouths are marked with only a small line or dot, either red or salmon pink. Their hair is embroidered in black, and only one Apostle has a brown embroidered hair line on top of the black one. Most likely the embroiderer meant to portray a figure of a fair complexion. It is worthwhile to note that none of the Apostles, in this case, is portrayed with white hair, i.e., of an elderly age. Usually in the Eastern iconographical prototypes, codices, Gospel illustrations, etc., the Apostles are represented throughout the Byzantine era with naturalistic features that follow specific iconographic rules, as presented in the most important work of the 17th century by Denys de Fourna, representing an iconographic variety of ages. (For the Apostles representations see [14] (pp. 298–299)). At first sight, the Apostles all look alike, but one will soon notice that each one has its own particular features and, thus, personality. Two of the twelve look younger, with oval faces traced in black, three wear a beard, and the rest have fat, almost round faces with no apparent neck.

Their feet portrayed in the traditional Coptic style are bare. This is a custom still preserved in Coptic ritual, where the priest is barefooted through the liturgy, symbolizing God's command to Moses to approach barefooted (in Exodus 3, 5). The same is seen in many icons and fresco depictions of Saints, for instance, in the icon of St. Antony and St. Paul in the Coptic Museum, as well as in the icon and the fresco of St. Thomas the Hermit.

Despite the fact that the Apostles seem to be in a static state, their feet show movement—only five of them have been staying still. Three, on the right side of the Crucifix, seem to walk towards Christ. The same goes for one Apostle in the other lateral composition, who also walks towards Christ, only in the opposite direction.

The three Apostles who wear a beard (perhaps meant to be older and wiser), two on the left and one on the right side of the scene of Figure 1, have their right hands raised in a gesture of testimony. Each one is holding an orb with an inscribed red cross in their left hand. They are believed to represent the Apostles Peter, James, and John (*John* XVII, 23), who are rendered in this same pose to symbolize the transfiguration (St. Catherine, Sinai) and the Ascension (The Rabbula Gospels). (For an example see [15] (p. 164 Icon in the Benaki Museum. Ascension, 17th century) and [2] (p. 77). In addition Matthew is selected as the 12th Apostle, in the place of Judas. Thus, replacing Juda in a symbolic depiction reminiscent of Juda's death, hanged by the tree, therefore being elevated and not touching the ground [16] (Figure 5; see also hanged Judas far right in Figure 1).

Figure 5. Does this elevated figure represent Matthew (Chapter B of the Acts of the Apostles) rather than representing Juda?

In any case, we believe that this unique representation of the Crucifixion with the twelve Apostles, which also involves the Ascension, is not only an unusual and unique compositional formula but allows us to presume that it represents Christ's death as a triumph over death, emphasizing, along with the other factors, the non-Chalcedonic origin of the embroidered panel.

The scene of the Crucifixion is framed by a row of small, embroidered crosses on top, alternatively displayed in black and white. On the right and left, the two vertical sides have, respectively, twelve crosses in white. On top of the scene lengthwise, there is a row of seventy crosses. We believe that the embroiderer has not embroidered the crosses merely by chance, but he or she must have chosen these respective numbers on purpose, trying to emphasize their symbolic meaning.

If we are on the right track, the twelve crosses on the vertical sides are referring to the twelve Apostles, the number twelve being the symbolic representation of the twelve tribes of Israel (*Matthew* XVIII, 28 and *Luke* VI, 13).

As for the seventy crosses all along the top of the Crucifixion scene, they must represent the Seventy Apostles, disciples who were named by Christ himself during his lifetime to spread his Word all over the World (*Luke* X, 1).

On the other hand, little crosses which surround an icon or decorate a haical (central Coptic chapel with altar) door, a haical curtain, an iconostasis, or the ecclesiastical vestments on the ΠΙΒΑΛΛΙΝ, as found in the extensive works of A. J. Butler, Evelyn White, L. A. Hunt and others, are of typical Coptic style, used for the exorcism of Satan, as pointed out by Reverend Pola Amba Bishoi of the Coptic Church in Athens [17]. Some signs may have been added for ornamentation's sake.

4. The Inscription

Finally, it is the issue of the inscription, which is embroidered in black lengthwise, in one line, exactly under the scene. It is separated at intervals by a chrism embroidered also in black  (Figure 6).

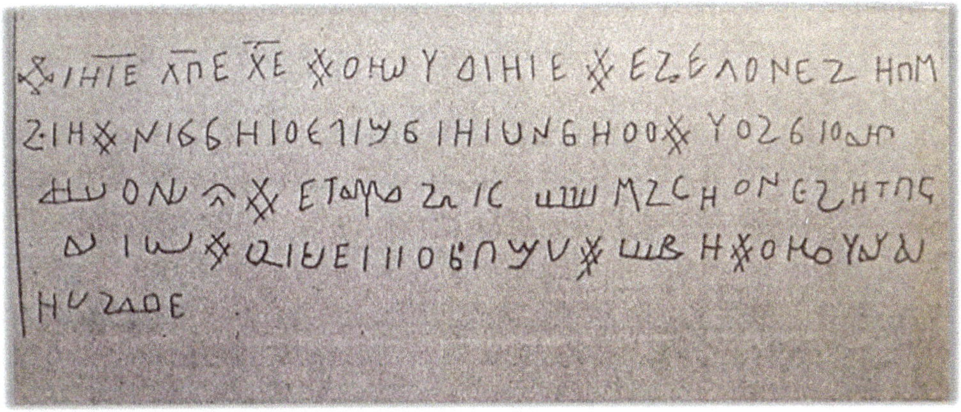

Figure 6. Unidentifiable inscription in Coptic, Greek, etc.; characters (Δοξαστικόν?) embroidered in black lengthwise, in one line, exactly under the scene, separated at intervals by a chrism.

The embroidered inscription is written in Greek and Coptic characters, while in some parts there is a variation of other letters, as is pointed out by Reverend Pola Amba Bishoi of being Syriac 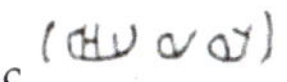. Tentatively, I recognized a galgolitic one (Q); some other letters are unidentifiable (may be Arabic), such as this one 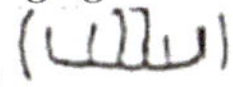. Nevertheless, the first three abbreviations at the beginning of the inscription give us a clue that they are of Greek meaning. The first abbreviation, although misspelled, is referring to the Monogram of Christ IHIE. The second one, , adding (E), could be read as (ΕΛΕΗΣΕ/have mercy), and the third (XE) with upper dash is deciphered with a great certitude as ΧΡΙΣΤΕ (Oh! Christ). On a 15th century unique icon of the Cretoitalian School, in the Byzantine Museum of Athens (T. 2638), is depicted a most rare representation of both the Crucifixion and the Resurrection in the East (Εις Άδου Κάθοδος = to the netherworld descendance) and western prototypes. The icon underneath has a two-row inscription of a supplicatory troparion and has been a donation. Thus, we are led to the conclusion that this inscription is a Δοξαστικον of either the holy Friday (ΔΟΞΑΣΤΙΚΟΝ of the Crucifixion; that is, the glorious type of hymn found in the Divine Services of the Eastern Orthodox Church) or a ΔΟΞΑΣΤΙΚΟΝ of the twelve Apostles (30th of June). A Coptic inscription with the same chrism, lozenge shaped at intervals, is also traced on a late Coptic tapestry fragment in the Louvre, that P. P. du Bourguet has described as a curtain or an altar hanging, from the Fayoum, 12th century [5] (p. 647). Some single signs remind us of Ge'ez writing, the script used as an alpha syllabary for several Afro-Asiatic and Nilo-Saharan languages of Ethiopia and Eritrea in the Horn of Africa. Its origin may have been an abjad (consonant-only alphabet) first used to write the Ge'ez language, which today is the liturgical language of the Ethiopian Orthodox. It is also a means of confirming dating [18]. The present textile object is unique, depicting Christian symbols, yet belongs to the wider investigation of Coptic embroideries. One of the biggest characteristics of Coptic textile motifs of the later period is dualistic expressions of Christianity and Islam, which were interpreted as virtue and evil, and civilized world and barbarous world [19], and, in some cases, use of bi- or multilingual aspects is observed [20].

5. Conclusions

The iconographic scene of the Crucifixion with the twelve Apostles in the Benaki Museum (Athens, Greece) is a most particular, mysterious, and unique representation of the Crucifixion. It is a hybrid with no resemblance to anything else. It has been proved iconographically that it has elements from both the East, the West, and the North. Stylistically in many details which are not apparent at first sight, one can trace Armenian connotations as, for instance, in the protuberant knot of Christ's loincloth (perizoma). In addition, one can notice a Syrian influence especially in the rendering of the Apostles' faces. The inscription in itself is a synthesis of letters from different alphabets. However, with its obvious syncretism, the particular personality of Coptic art prevails, with its abstract, conventional forms and its pronounced mysticism and symbolism, that, nonetheless, remains deeply Christian. As for its place of origin, our working hypothesis would assign it to the Low Egypt region of Fayoum. Having no iconographic or stylistic parallels of the Crucifixion here examined, it is practically impossible to gain any insight in its earlier history and in its origin, dated at the 13th–14th century.

Funding: This research received no external funding.

Acknowledgments: I would like to extend warm thanks to M. Rassart-Debergh and D. Johnson of the UCL Louvain-la-Neuve, late A. Delivorrias of the University of Athens, Maria Theoharis of the University of Bari, Marie-Helene Rutschowskaya, Chief Curator at the Louvre, R. Pola Amba Bishoi of the Coptic Church of Athens, and John Meimaris of the Byzantine Department of the National Hellenic Research Foundation for valuable comments and fruitful discussion in written form. Many thanks to the two anonymous referees and I. Liritzis for critical comments.

Conflicts of Interest: The authors declare no conflict of interest.

References

1. Meinardus, O. The collection of Coptica in the QASR of the Monastery of St. Antony. *Bull. Société Archéologie Copte* **1980**, *18*, 272.
2. Delivorrias, A. *Guide to the Benaki Museum*; Benaki Museum: Athens, Greece, 1980; p. 34.
3. Millet, G. *Recherches sur L'iconographie de L'évangile*; Fontemoing et Cie: Paris, France, 1916.
4. Schiller, G. *Iconography of Christian Art*; New York Graphic Society: New York, NY, USA, 1972; Volume 2, figs. 323–324.
5. du Bourguet, P.R.P. *Catalogues des Étoffes Coptes I*; Musée National du Louvre: Paris, France, 1964.
6. Lamm, C.J. *Cotton in Medieval Textiles in the Near East*; P. Geuthne: Paris, France, 1937.
7. Thomson, D. Cotton double cloths and embroidered and brocaded linen fabrics from tenth to fourteenth century Egypt: Their relation to traditional Coptic and contemporary Islamic Style. *Bull. Liaison CIETA* **1985**, *61*, 1–2.
8. Paulus, B. *Pascasius Radbertus De Corpore et Sanguine Domini: Cum Appendice Epistola ad Fredugardum*; Corpus Christianorum, Continuatio Mediaeualis; Turnholti, Brepols: Turnhout, Belgium, 1969; Volume 16.
9. Child, H.; Colles, D. *Christian Symbols—Ancient and Modern*; Bell and Hyman: London, UK, 1960.

10. Soteriou, G.A.; Soteriou, M. *Eikones tes Mones Sina/Icones du Mont Sinai;* Collection of the French Institute of Athens: Athens, Greece, 1956; Volume 2.
11. Weitzmann, K. *Icons from Southeastern Europe and Sinai;* Thames and Hudson: London, UK, 1968.
12. Soteriou, G.A. *The Christ on Art;* N. Tzakas: Athens, Greece, 1914.
13. Butler, F.J. *The Ancient Coptic Churches of Egypt;* Gorgias Press: Oxford, UK, 1884; Volume 2.
14. Dionisius from Fourna (Διονυσίου εκ Φουρνά). *Interpretation of Art Painting;* French Translation Denys de Fourna, Manuel D'iconographie Chretienne par A; Papadopoulos-Kerameus: Saint Petersbourg, Russia, 1909; pp. 298–299.
15. Grabar, A. *La Peinture Byzantine;* Bibliothéque Laurentienne: Genéve/Florence, Italy, 1953; p. 164.
16. Loverance, R. *Byzantium;* British Museum Press: London, UK, 1988.
17. White, E. *The History of the Monasteries of Nitria and the Scetis;* 1931; Metropolitan Museum of Art, Egyptian Expedition: New York, NY, USA, 1975; Volume 8.
18. Miqdadi, R.H. Abjad numerals as an absolute dating method: Forts from al-Ain, UAE. *Mediterr. Archaeol. Archaeom.* **2020**, *20*, 273–289. [CrossRef]
19. Han, J. A study on the characteristics of the designs on Coptic textiles of Ancient Egypt. *J. Fash. Bus.* **2011**, *15*, 112–124.
20. Richter, T.S. Greek, Coptic and the 'language of the Hijra': The rise and decline of the Coptic language in late antique and medieval Egypt. In *From Hellenism to Islam: Cultural and Linguistic Change in the Roman Near East;* Cotton, H.M., Hoyland, R.G., Price, J.J., Wasserstein, D.J., Eds.; Cambridge University Press: Cambridge, UK, 2009; pp. 401–446.

heritage

Article

Monitoring Coptic Masonry Affected by Clay Minerals and Microorganisms at the Church of Virgin Mary, Wadi El-Natrun (Egypt)

Abubakr Moussa [1,*] and Mahmoud Roshdy [2]

1 Department of Conservation, Faculty of Archaeology, Cairo University, Giza 12613, Egypt
2 Department of Islamic Antiquities, Faculty of Archaeology, Cairo University, Giza 12613, Egypt; mahmoudsalem@cu.edu.eg
* Correspondence: dr_abubakr@cu.edu.eg; Tel.: +20-(10)-9666-4411

Abstract: This paper focuses on the role played by the clay minerals and microorganisms in the deterioration process of Coptic architecture units at the church of Virgin Mary, Wadi El-Natrun region. For this purpose building materials (mainly mortars and plasters) from the studied church were examined using X-ray diffraction (XRD), X-ray fluorescence (XRF) and scanning electron microscope with energy dispersive spectroscopy (SEM-EDS); in order to identify their composition and were investigated petro-graphically to determine the real response of the masonry structure to the deformation imposed at the endogenous factors. Wall gypsum mortars in the church contain halloysite as a dominant clay mineral while plaster is clay free; concerning microorganisms, the fungal flora *Aspergillus glaucus* represent the most dominant fungi constituting (22.22%), *Aspergillus flavus*, *Aspergillus flavus*, *Aspergillus fumigatus*, *Aspergillus occhraceus*, and *Aspergillus caudidus* were also isolated.

Keywords: coptic; masonry; clay minerals; microorganisms; XRD; XRF; Wadi El-Natrun

Citation: Moussa, A.; Roshdy, M. Monitoring Coptic Masonry Affected by Clay Minerals and Microorganisms at the Church of Virgin Mary, Wadi El-Natrun (Egypt). *Heritage* **2021**, *4*, 4056–4067. https://doi.org/10.3390/heritage4040223

Academic Editors: Christofilis Maggidis, Nikolaos Lazaridis, Omar Abdel-Kareem and Robin Skeates

Received: 23 August 2021
Accepted: 26 October 2021
Published: 29 October 2021

1. Introduction

Wadi El Natrun, is a sandy depression located in the western part of the Nile delta, latitude 30°17′ and 30°38′ N, longitude 30°02′ and 30°30′ E. It is directed NW-SE, about 22 m under sea level. In the region many salty alkaline lakes are found half-way between Cairo and Alexandria. Wadi El-Natrun is one of the oldest centers of monastic settlements in Egypt, Figure 1.

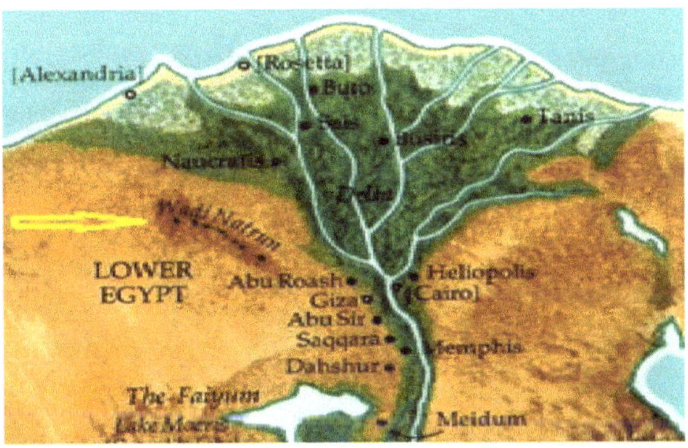

Figure 1. A map showing the location of Wadi El-Natrun area (http://www.lib.utexas.edu/maps/, accessed on 18 October 2021).

Deir "Monastery" Al-Surian, where the studied church is located, was called Al-Surian because it was owned by the Syrians. Its real name is Monastery of Saint Mary. It was constructed in the first half of the (1st AD) century in the place of the Monastery of St. John the Short and was listed in the history of monasteries. It was a secondary monastery that emerged as an independent unit in the second half of the 3rd AD century. In (817 AD), the monastery—as well as others—was destroyed, and the Syrian monks rebuilt it. It did not have a wall until the late 9th AD century. The Monastery of Saint Mary (Al-Surian) in Wadi El-Natrun is rectangular and measures 150 m long and 45 m wide, Figure 2.

Figure 2. A general view of Deir "Monastery" Al-Surian, Wadi El-Natrun area.

It is divided into two parts: the western and the eastern part. The western part up to the west of the gate contains palaces. It was built in (850 AD) before the building of the wall. In contrast, the eastern part is the largest and contains the Church of Virgin Mary adjunct to the northern wall, a mill in front of the church, and a garden of palm trees in the east. The church is a rectangular church, where its horizontal projection is a rectangle of 24 m long and 10 m wide from inside. It has two covered roofs: the northern and the southern that was separated when building the wall of the monastery. The eastern part of the church is divided into three altars bordered from the west by the choir across the church. The church also contains porticos: a middle portico, side porticos, and the entrance of the portico. The southern entrance was separated when building the wall. The current northern entrance dates back to (980–1000 AD) Figure 3.

It is a high square building that can be accessed through a path with a high pointed arch in the wall and arched doorstep as well. From the inside, the projection is square. On the western side, there is a door with an arched doorstep and double entrance. Moreover, the eastern wall has a similar entrance to the "Presbyterian Church". The dome of the entrance is pointed and based on a square that became an octagon by four corner squinches, then a circle by eight small Turkish triangles. The middle portico is covered by a 9 m high vault. Although it is one block, it is divided by a 1.25 m high section of buildings. Consequently, the eastern section represents an archway used as a choir. It is known as the Laqan Basin, which is shallow and filled with water on the Maundy Thursday that commemorates the Washing of the Feet (Maundy) of Jesus Christ with the Apostles. To the west of the middle portico, there is a portico with an arch holding half dome based on wooden beams on the western wall. This dome was built of small stones. Side porticos are covered by half-cylindrical vaults of 4.5 m high about half of the middle portico. Their projection is 1.50 m wide. In the eastern ending, the porticos are blocked with no doors to the choir. The eastern archway of each portico is separated by a short section, which completes the section separating the portico that adjoins to the external choir, finally the separated section is covered by a vault. Above the southern portico, there are upper windows with a set of bars that form octagons and small triangles with glass pieces. The choir is covered by an 11 m high dome. The only entrance to the middle portico to the choir is an opening with a semi-circular doorstep. The side columns have a door that dates back to the (10th AD) century. The middle part of the choir facing the middle altar is topped by a pointed dome based on two wide pointed arches. The dome is supported by a half dome on the northern and southern endings of the choir. The middle altar is covered by a high dome of 10 m high based on an octagon. Its ground is higher than the choir by three wide steps. There are two squinches on the northern and southern walls of the middle altar. These two walls have entrances leading to the side altars. Moreover, the doors of the altars are fixed on wooden door jams similar to the doors of the choir. In other words, they are opened from the middle and based from the back on bronze hinges. Each door has three shutters, each of which contains seven panels. The doors date back to the (8th AD) century. The altar is a hollow square of buildings and covered by a square-circular tile (i.e., a side is square and the other is semicircular, unlike the churches of the desert). It is made of marble or a black marble-like stone block. Two entrances in the form of a narrow room that is a bit longer than the middle altar with circular doorsteps lead from the choir to the side altars. This pattern was common in the (9th AD) century. Each room has a small altar and an eastern squinch. Concerning side altars, each side altar is covered by a semi-cylindrical vault that is 4.5 m high, Figure 4, [1–6].

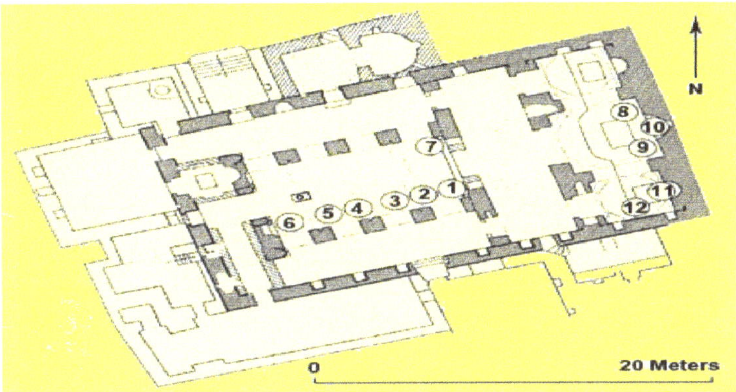

Figure 3. A plan of the Church of the Virgin showing: **1.** Annunciation. **2.** Visitation. **3.** Nativity. **4.** Baptism of Christ. **5.** Tow jars on Table. **6.** Entry to Jerusalem. **7.** Pentecost. **8.** Sacrifice of Isaac. **9.** Meeting of Abraham and Melchizedec. **10.** Christ and the Holly Virgin enthroned. **11.** Saints. **12.** Saints.

Figure 4. (**Left**): The church of the Virgin, Deir al-Surian, Wadi El-Natrun. (**Right**): A 3D model of the church of the Virgin, Deir al-Surian, Wadi El-Natrun.

Atmospheric weathering of quarried aggregates is capable to produce clay minerals; the swelling of clay minerals develops sufficient pressure to break up the building materials [7]. An important role is played by the microscopic fungi in atmosphere-exposed environment. The problem of longevity increase of archaeological buildings can be solved only with deep understanding of the mechanisms of deterioration. As it is known from other inorganic materials (stones, mortars, plasters and pigments) a special attention should be given to interactions of materials with microorganisms and to the processes of their microbial degradation [8]. The current study aims essentially to assess the effect of clay minerals and microorganisms in the deterioration cycle of Coptic architectural units at the Church of Virgin Mary, Wadi El-Natrun, using chemical and physical analyses; this may be very useful in correctly analyzing the development of damage picture. By this way the results of the present study may be extrapolated to be used in similar cases suffering from the same problems.

2. Materials and Methods

X-ray diffraction (XRD) analysis was performed using a Philips (PW1840) diffractometer with Ni-filtered Cu-Kα radiation. The samples were scanned over the 5–70° 2Θ intervals at a scanning speed of 1.2° min^{-1} A° quantitative estimation of the abundance of the mineral phases was derived from the XRD data using the intensity of certain reflections and external standard mixtures of minerals compared to the JCPDS standards of 1967 [9]. The µ-XRF analyses used a diffractometer type: SPECTRO, Model: COPRA "Compact Portable Roentgen Analyzer", potential acceleration: 35 kV, lamp stream: 0.9 mA, analysis time: 300 s. for the SEM-EDS analyses; samples were coated by the JEOL JEE-4X High vacuum evacuator with carbon coating (carbon rods). The SEM is the JEOL JSM-840A, with an EDS analytical attached of OXFORD instruments ISIS, working distance (WD) 20 mm, and current 1 mA, accelerate voltage 20 kV. For the clay mineralogy study, grain size determination of the material and textural classification was performed on each sample following the Folk method [10]. A 20 g split of each sample was subjected to the following chemical treatments [11]: 1 N sodium acetate-acetic acid buffer solution with pH = 5.0 for carbonate removal; 30% H_2O_2, for organic matter and Mn-oxides removal; 0.3 M sodium citrate and 1 M sodium bicarbonate buffer solution with pH = 7.3 to which 1 g increments (up to 3 g) of sodium dithionite were periodically added to remove free Fe-oxides and Fe-Al-hydroxides. Fractions having dimension less than 2 microns were separated into other three different fractions (2–1, 1–0.25 and <0.25 µm) by an international equipment company (IEC) centrifuge. The sample fractions were dried in room temperature. Subsequently, random and oriented mounts were prepared for XRD analysis. All the oriented mounts were reanalyzed after treatment with an ethylene-glycol solution to distinguish the expandable mineral phases. Some were heated for 2 h at 550 °C for chlorite detection [12]. Semi-quantitative estimates of the mineral abundance based on the peak area of the oriented mounts, were made from the XRD data, using the method described in [13]; the study was carried out in the School of Geology, Aristotle University of Thessaloniki, Greece. Concerning the biological study Czapeck's medium [14,15], was used for isolating and growing microorganisms. Three replicate plates were prepared for each sample. After 1 week incubation period at 28 °C, fungal colonies were purified using Martin medium, purified fungal isolates were identified using PDA medium, which was used as a slant stock cultures for fungi according to [15–17]. The study was achieved in the School of Veterinary, Aristotle University of Thessaloniki, Greece.

3. Results

A mortar sample from the western wall inside the church was analyzed using the XRD method. The sample is composed of 51% calcite ($CaCO_3$), 32% gypsum ($CaSO_42H_2O$), 5% anhydrite ($CaSO_4$) and 1% plagioclase (mainly albite), as seen in Figure 5 (left). The same results were emphasized by the SEM-EDS analysis, calcium, sulfur, oxygen and carbon are the main sample components, as shown in Figure 6a. The presence of high amount of gypsum in the sample was also assured by the use of μ-XRF, as seen in Figure 7a.

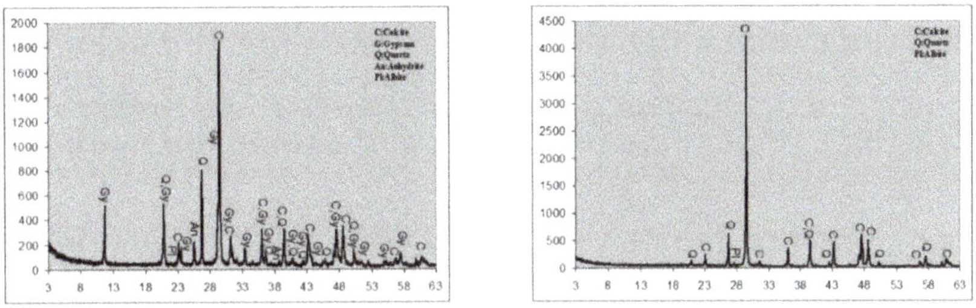

Figure 5. Powder X-ray diffraction patterns from the church of the Virgin showing: (**Left**) The mortar sample which composes mainly of calcite and gypsum. (**Right**) The plaster sample which composes mainly of calcite (*X*-axis = 2Θ, *Y*-axis = Counts).

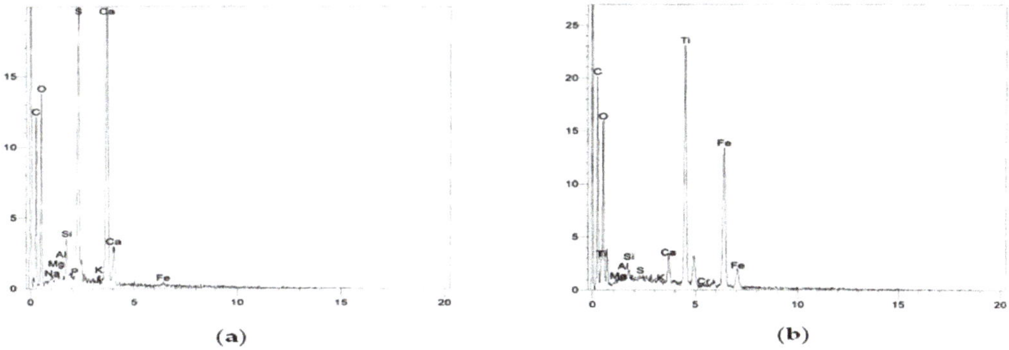

Figure 6. Energy dispersive spectrum from the church of the Virgin showing: (**a**) A mortar sample from the west wall inside the church of the Virgin; (**b**) The plaster sample from the west wall inside the church of the Virgin, (*X*-axis = E/keV, *Y*-axis = Cps.).

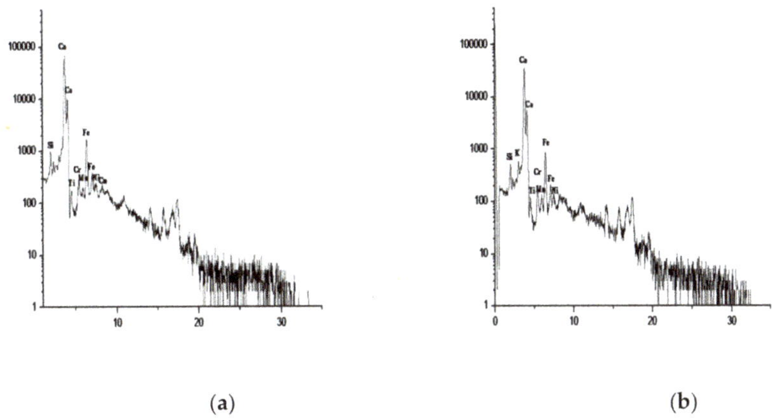

Figure 7. μ-XRF analysis patterns from the church of the Virgin showing: (**a**) A mortar sample from the west wall inside the church of the Virgin, (**b**) The plaster sample from the west wall inside the church of the Virgin, (*X*-axis = E/keV, *Y*-axis = Imp.).

Concerning plaster, only one sample was available. This sample was taken from the surficial layer in an unpainted part of the western wall inside the church. The plaster layer from which the sample was derived has been added in 1781/2 during a re-plastering process [1–6]. According to the XRD analyses, Figure 5 (right), the plaster is lime plaster, and is composed of 92% calcite ($CaCO_3$), 7% quartz (SiO_2) and 1% plagioclase (mainly albite). By the analyses of both μ-XRF (Figure 6b) and SEM (Figure 7b), the sample is lime, composed basically of calcite and quartz. Tables 1 and 2 highlight the major and minor elements detected in the studied mortar and plaster samples.

Table 1. The chemistry of the studied mortar sample from the church of the Virgin, Wadi El Natrun.

Element %		Atomic %	Compound %	
Na	3.58	3.54	Na_2O	4.83
Mg	5.75	5.37	MgO	9.54
Al	4.17	3.51	Al_2O_3	7.89
Si	21.57	17.44	SiO_2	46.14
S	0.24	0.17	SO_3	0.60
Cl	12.32	7.89		0.00
K	1.07	0.62	K_2O	1.29
Ca	9.40	5.33	CaO	13.15
Fe	3.30	1.34	FeO	4.24
O	38.59	54.78	-	-
Total	100%	100%	-	87.68%

Table 2. The chemistry of the studied plaster sample from the church of the Virgin, Wadi El Natrun.

Element %		Atomic %	Compound %	
Na	39.09	43.51	Na_2O	52.69
Mg	0.06	0.07	MgO	0.10
Al	0.11	0.11	Al_2O_3	0.21
Si	0.26	0.24	SiO_2	0.56
S	−0.06	−0.05	SO_3	−0.14
Cl	45.28	32.68		0.00
K	0.12	0.08	K_2O	0.14
Ca	0.66	0.42	CaO	0.92
Fe	0.18	0.08	FeO	0.23
O	14.29	22.86	-	-
Total	100%	100%	-	54.72%

3.1. Clay Mineralogy

The mortar sample that was taken from the western wall of the church of the Virgin in Wadi El Natrun was treated according to the same methodology mentioned before (materials and methods), the sample, in the oriented phase, was found to contain only a layer of halloysite clay mineral at 11.6 A°, while the peak appears at 21 A° stands for Quartz (SiO_2), Figure 8a. In the glycolated phase the same layer "halloysite" was detected at 11.3 A°, while the peak appears at 21 A° stands for quartz (SiO_2) (Figure 8b). After heating, the halloysite layer disappeared and no other clay minerals were detected in the sample in that phase, while the studied plaster sample did not show any residues of any kind of clay minerals at the three phases (oriented, glycolated, heated).

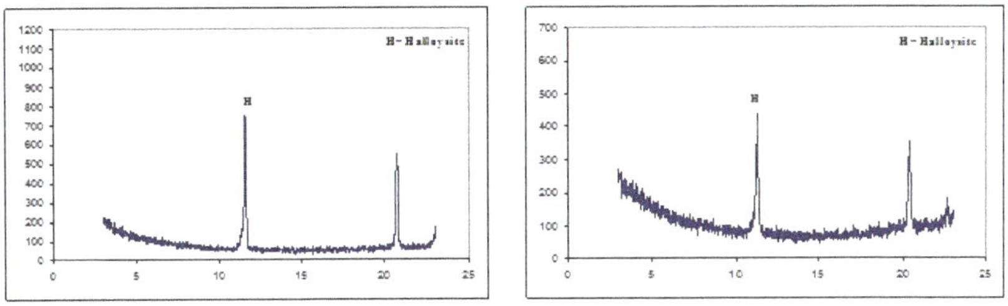

(**a**) (**b**)

Figure 8. XRD patterns of the studied mortar sample from the church of the Virgin showing: (**a**) Oriented phase; (**b**) Glycolated phase (*X*-axis = 2Θ, *Y*-axis = Counts).

3.2. Biological Study

The identification process which was achieved according to [18] proved that; the fungal flora *Aspergillus glaucus* represent the most dominant fungi constituting (22.22%) of the total fungal isolates from the fresco paintings in Wadi El Natrun and it was found to attack green, red pigments, plasters and it was also isolated from the air of the church of the Virgin. Other Aspergillus species identified were: *Aspergillus flavus* (11.11%) which was found to attack black and white pigments. *Aspergillus flavus* + *Aspergillus niger* which were found to attack white, green and orange pigments in the fresco paintings of the church. *Aspergillus fumigatus* (16.66%) which has been identified in the isolates taken from the plasters applied on walls and columns in the church. The *Aspergillus occhraceus* group was found only on the yellow pigment in the church. *Aspergillus caudidus* was identified in Wadi El Natrun fresco paintings and was found to grow on yellow and black pigments. Other genera were: *Mucor* spp. which was found only in the internal air of the church of the Virgin. The fungus *Geotrichum* spp. has been found to grow only on the red pigment in the church. *Geotrichum* spp. and *Aspergillus fumigatus* were identified only in the isolates taken from the un-plastered walls of the church (Figures 9a,b, 10a,b and 11a,b and Table 3).

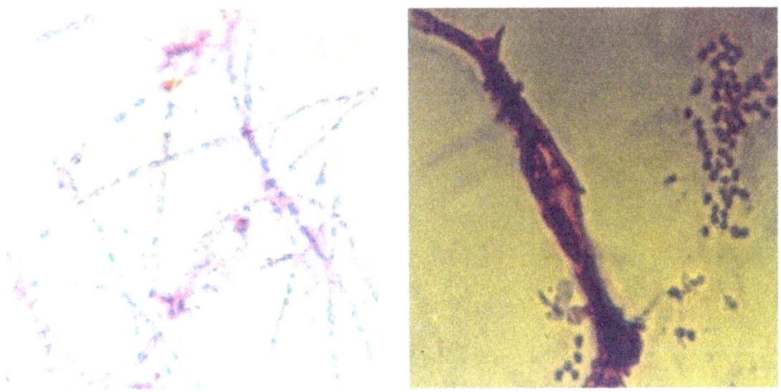

(**a**) (**b**)

Figure 9. The fungal flora *Aspergillus caudidus* (**a**) under microscope (**b**) (un-plastered walls of the church—1500×).

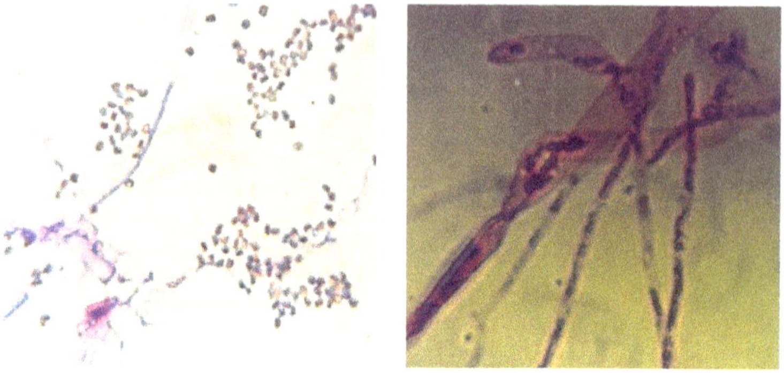

(**a**) (**b**)

Figure 10. The fungal flora *Mocur spp.* (**a**) under microscope (**b**) (the internal air of the church—1500×).

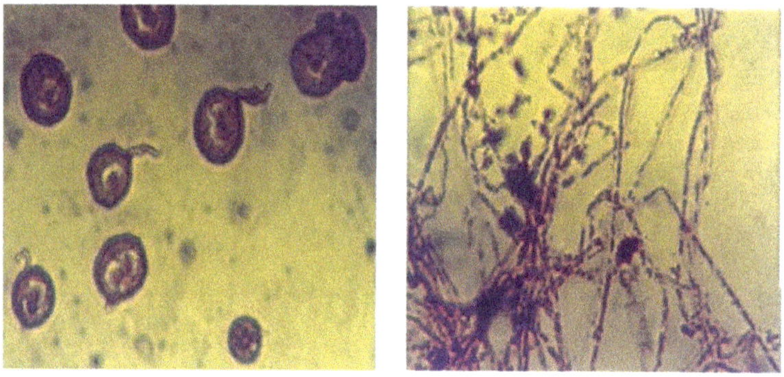

(**a**) (**b**)

Figure 11. *Aspergillus flavus* (**a**) and *Aspergillus niger* (**b**) under microscope (1500×).

Table 3. The fungal flora isolated from the church of the Virgin—Wadi El Natrun.

Isolated Fungal Species	No. of Isolates	Percentage %
Aspergillus flavus group	2	11.11
Aspergillus flavus + *Aspergillus niger*	3	16.66
Aspergillus fumigatus	3	16.66
Aspergillus glaucus	4	22.22
Aspergillus occhraceus group	1	5.55
Mucor spp.	1	5.55
Aspergillus caudidus	2	11.11
Geotrichum spp.	1	5.55
Geotrichum spp. + *Aspergillus fumigatus*	1	5.55
Total isolates	18	100%

4. Discussion

Gypsum and anhydrite were found together in the mortar sample; both gypsum and anhydrite are rock-forming minerals, the formation is sometimes referred to as gyprock. Some of these deposits can be hundreds of meters thick. In dry climates, anhydrite deposits have remained the same for over millennia. However, in the presence of groundwater they can turn into gypsum to some 15 to 40 m depth, this depth increasing two-fold in humid climates. The hydration of anhydrite (generally called gypsification) and the volume expansion associated with this reaction has been considered the cause for the observed powdering of the surface of anhydrite rock into gypsum [19]. Mortars and plasters containing gypsum have a tendency to deterioration because of their solubility and their conversion to anhydrite [20,21]. Plaster in the church was applied in five layers [22,23], the analysis results evidenced the complete change into lime in Egypt after millennia of gypsum and gypsum–lime mixtures. Mortars and plasters consisting primarily of lime and sand has been used as an integral part of masonry structures for thousands of years, this basic formulation remained unchanged until the advent of Portland cement, the latter is sensitive to the attack of sulphates and sea water since calcium aluminates react with (SO_4), forming ettringite (a gypsum salt). Moreover, cement mortars interact with atmospheric (SO_2) forming gypsum and ettringite [24,25].

Most clay silicates occur as thin plates, but halloysite often occurs as tubular or spherical particles [26]. Halloysite is usually found in soils formed from volcanic deposits, particularly volcanic ash and glass [27]. Halloysite is a polytype (hydrated phase) of kaolinite ($Al_2(OH)_4[Si_2O_5]$) and differs mainly in the morphology of the crystals, which are curved or rolled, this leads to structural disorder, particularly in layer stacking [28]. As halloysite is a layer-structured mineral, most of its surfaces are internal (referred to as inner surface). When low molecular weight organic compounds react with these inner surfaces, an expansion occurs along the c-axis [29]. Halloysite is an indicator for arid climate conditions; it is frequently formed under wet conditions in the rock soil interface zone, in hydrothermal springs and subtropical weathering of volcanic rocks, in cracks and around weathering of granite boulders. The presence of halloysite in soils and building materials is associated with movement of surface water to ground water [30,31]. Mechanically, the clay mineral halloysite ($Al_2Si_2O_5(OH)_4.2H_2O$) absorbs the soluble salts resulting—through swelling—in endogenous pressure on the stone blocks and the plaster layer leading to wall bulking, plaster falling and cracking [23].

The degradation of natural materials depends on the climatic conditions and human activity. In a monument, all building materials are not subjected to the same weathering processes because of their intrinsic properties (mineralogical and chemical composition, petrophysical and mechanical properties) and their position in a monument [32]. Building materials kept in high controlled environments such as museums are also susceptible to the effect of bio-deterioration—especially the effects of fungal and bacterial species. Biocorrosion of building materials can occur through the action of microorganisms leading to stone dissolution and formation of different salts; these salts can be precipitated within the pores of the stone, and upon recrystallization, exert considerable stress in the pore walls [33]. Fungal and bacterial species uses the copper based-pigments as a nutrient, a main source for carbon and energy leading to the chromatic alteration of these pigments [34], the organisms associated with biodeterioration processes on buildings grow, as biofilms, containing a complex community of: microorganisms, particulates and a high percentage of water forming under suitable moisture conditions, a gelatinous film [35]. Gettens and coworkers were among the first to point out, in 1941, that paintings could be defaced or destroyed by the growth of those small, parasitical plants commonly called "mold" or "mildew" [36]. All building materials are porous and the porosity of building material along with ingress of moisture and other harmful chemicals such as acids, chlorides and sulfates affect the material and seriously reduce their strength and service life [37], bio-corrosion can occur through the excretion of organic acids by chemoorganotrophic microorganisms. These acids can also chelate metal cations (e.g., Al, Ca, Fe, Mg, Mn and Si) from minerals to form stable complexes [33]. The presence of significant amounts of carbonate compounds (e.g., >3% *w/w* $CaCO_3$) in calcareous stones or lime mortars, results in the buffering of biogenic metabolic products producing a constant suitable pH-milieu for the growth of microorganisms [38]. Moreover, these microbes are also capable to precipitate $CaCO_3$ via hydrolysis of urea [39]. These species are responsible for the biologically induced corrosion of stone and other building materials in the archaeological sites. Sometimes, death and lyses of these species provide the organic substrates necessary for the growth of heterotrophic species, which are responsible for the colored stains present on the building surface as well as the mechanical damage observed, such as the detachment of portions of the painted layers. Microorganisms which grow on the surface of wall paintings might discolor the painting not only through their own pigments but also by excreting metabolic products which cause discoloration of the paint or change its consistency. In addition, mycelia of fungi can penetrate into the painting and its grounding, resulting in the mechanical destruction of the cultural heritage [40]; generally, two common types of aesthetic damage can be observed: white efflorescence and green-to-black stains. The same deterioration symptoms are observed in our case study. Concerning biological control, recent studies demonstrated that heating the infected area of the building surface to 70 °C for a few minutes by a microwave system represents an effective and sustainable method to devitalize biofilms that grow on the surface of ancient buildings, yielding the same successful devitalization obtained by an abundant biocide application, but avoiding any dispersal of toxic residues in the environment [41].

5. Conclusions

A certain role is played by the clay minerals and microorganisms in the deterioration process of Coptic architecture units at the church of Virgin Mary, Wadi El-Natrun region. Wall gypsum mortars in the church contain halloysite as a dominant clay mineral while plaster is clay free. Concerning microorganisms, the fungal flora *Aspergillus glaucus* represent the most dominant fungi constituting (22.22%), *Aspergillus flavus, Aspergillus flavus, Aspergillus fumigatus, Aspergillus occhraceus,* and *Aspergillus caudidus* were also isolated, the findings of this study are important in correctly analyzing the development of damage cycle and is a rational definition of the restoration work necessary. By this way, the results of the present study may be extrapolated to be used in similar cases suffering from the same problems the mortars proposed for restoration should ensure compatibility with the historic structure and also feature some improved characteristics, for a greater stability and mostly sustainability. These concern, basically, a higher, but not excessive, compressive strength, an adequate tensile strength, elasticity and an enhanced hydraulic character. Concerning biodeterioration, our laboratory experimental work in similar cases assured that the optimum biocide for both fungi and bacteria is marjoram oil (*Origanum vulgare* L. "Lamiaceae"), this oil possesses strong antibacterial and antifungal activities, it can be used as a solution composes of (2 cm^3 marjoram oil, 2 cm^3 acetone and 100 cm^3 distilled water). The biocide can be applied using the spray technique and has no side effects related to the aesthetic or chemical composition of the building materials.

Author Contributions: Conceptualization, A.M. and M.R.; Data curation, A.M. and M.R.; Investigation, A.M.; Methodology, A.M.; Writing–original draft, M.R.; Writing–review & editing, A.M. All authors have read and agreed to the published version of the manuscript.

Funding: This research received no external funding.

Institutional Review Board Statement: Not applicable.

Informed Consent Statement: Not applicable.

Data Availability Statement: Data available upon request from the authors.

Acknowledgments: We would like to thank Karel Innemie from Leiden University (Netherlands, head of the Dutch archaeological mission at Wadi El-Natrun, who provided us data and some of the figures inserted in this study.

Conflicts of Interest: The authors declare no conflict of interest.

References

1. Parandowska, E. Results of the Recent Restoration Campaigns (1995–2000) at Dayr al-Suryan. In *Christianity and Monasticism in Wadi Al-Natrun: Essays from the 2002 International Symposium of the Saint Mark Foundation and the Saint Shenouda the Archimandrite Coptic Society*; The American University in Cairo: Cairo, Egypt, 2009; Volume 3, p. 272.
2. Innemie, K.C. Recent Discoveries of Wall-Paintings in Deir Al-Surian. *Hugoye: J. Syr. Stud.* **1998**, *1*, 1–14.
3. Innemie, K.C. Deir al-Surian (Egypt): Its Wall-paintings, Wall-texts, and Manuscripts. *Hugoye J. Syr. Stud.* **1999**, *2*, 7–15.
4. Innemie, K.C. Deir al-Surian (Egypt): New Discoveries of January 2000. *Hugoye J. Syr. Stud.* **2000**, *3*, 1–7.
5. Innemie, K.C. Deir al-Surian (Egypt): Conservation work of autumn 2000. *Hugoye J. Syr. Stud.* **2001**, *4*, 1–10.
6. Innemie, K.C.; Van Rompay, L. Deir al-Surian (Egypt): New Discoveries of 2001–2002. *Hugoye J. Syr. Stud.* **2002**, *5*, 1–12.
7. Riemer, W. Degrading Igneous Rock-Aspects and Identification of the Geotechnical Problem. In Proceedings of the International Symposium on Industrial Minerals and Building Stones, Istanbul, Turkey, 15–18 September 2003; pp. 841–848.
8. Kondratyeva, I.A.; Gorbushina, A.A.; Boikova, A.I. Biodeterioration of Construction Materials. *Glass Phys. Chem.* **2006**, *32*, 254–256.
9. JCPDS. *Jaint Committee on Powder Diffraction Standards, Index to the Powder Diffraction File*; American Society for Testing and Materials: West Conshohocken, PA, USA, 1967.
10. Folk, R.L. *Petrology of Sedimentary Rocks*; Hemphill's Publishing Company: Austin, TX, USA, 1968.
11. Jackson, M.L. *Soil Chemical Analysis—Advanced Course*; 11th print; Self-published: Madison, WI, USA, 1974.
12. Pehlivanoglou, K.; Tsirambides, A.; Trontsios, G. Origin and Distribution of Clay Minerals in the Alexandroupolis Gulf, Aegean Sea, Greece, Estuarine. *Coast. Shelf Sci.* **2000**, *51*, 61–73. [CrossRef]
13. Biscaye, P.E. Mineralogy and Sedimentation of Recent Deep Sea Clay in the Atlantic Ocean and Adjacent Seas and Oceans. *Geol. Soc. Am. Bull.* **1965**, *76*, 803–831. [CrossRef]
14. Booth, C. *Methods in Microbiology*, 4th ed.; Academic Press: New York, NY, USA, 1971.
15. Stevens, R.B. *Mycology Guide Book*; Mycology Guide Book Committee, Mycology Society of American Universities, Washington Press: Seattle, WA, USA, 1981.
16. Barnett, H.L.; Hunter, B.B. *Illustrated Genera of Imperfect Fungi*, 3rd ed.; Burgess Publication Convention: Minneapolis, MN, USA, 1972.
17. Domsch, K.H.; Gams, W.; Anderson, T.H. *Compendium of Soil Fungi*; Academic Press: London, UK, 1980; Volumes 1 and 2.
18. Ellis, D.; Davis, S.; Alexiou, R.; Bartely, R. *Description of Medical Fungi*, 2nd ed.; University of Adelaide: Adelaide, Australia, 2007; Volume 2.
19. Lazzarini, L.; Laurenzi, T.M. *Il Restauro della Pietra (The Restoration of Stone)*; No. 22; Cedam: Padua, Italy, 1986.
20. Kamel, A.M.A. Dehydration of Gypsum Component of Plasters and Stuccos in Some Egyptian Archaeological Buildings and Evaluation of K_2SO_4 Activator as a Consolidant. *Sci. Cult.* **2019**, *5*, 49–59. [CrossRef]
21. Theologitis, A.; Kapridaki, C.; Kallithrakas-Kontos, N.; Maravelaki-Kalaitzaki, P.; Fotiou, A. Mortar and Plaster Analysis as a Directive to the Design of Compatible Restoration Materials in Frangokastello (Crete). *Mediterr. Archaeol. Archaeom.* **2021**, *21*, 109–120. [CrossRef]
22. Moussa, A. Assessing the Decay Agents of Wall Paintings in Al Qurna and Wadi El-Natrun Regions-Egypt. Ph.D. Thesis, Geology Department, Faculty of Science, Aristotle University of Thessaloniki, Thessaloniki, Greece, 2007.
23. Moussa, A.; Kantiranis, N.; Voudouris, K.; Stratis, J.; Ali, M.; Christaras, B. Diagnosis of Weathered Coptic Wall Paintings in the Wadi El Natrun Region, Egypt. *J. Cult. Herit.* **2009**, *10*, 152–157. [CrossRef]
24. Robert, C.M.F.; Speweik, J.P. Repointing Mortar Joints in Historic Masonry Buildings. *Preserv. Brief* **1998**, *2*, 1–16.
25. Toumbakari, E.E. Lime-Pozzolan-Cement Grouts and Their Structural Effects on Composite Masonry Walls. Ph.D. Thesis, Katholieke Universiteit Leuven, Faculteit Toegepaste Wetenschappen, Departement Burgerlijke Bouwkunde, Laboratorium Reyntjens, Heverlee, Belgium, 2002.
26. Schulze, D.G. Clay Minerals. *Encycl. Soils Environ.* **2005**, *2005*, 246–254.
27. Homburg, J.A. Archeology in Relation to Soils, Earth Systems and Environmental Sciences. *Encycl. Soils Environ.* **2005**, *2005*, 95–102.
28. Carrillo, A.M.; Carriazo, J.G. Cu and Co Oxides Supported on Halloysite for the Total Oxidation of Toluene. *Appl. Catal. B Environ.* **2015**, *164*, 443–452. [CrossRef]
29. Horváth, E.; Kristóf, J.; Frost, R.L.; Rédey, Á.; Vágvölgyi, V.; Cseh, T. Hydrazine-Hydrate Intercalated Halloysite under Controlled-Rate Thermal Analysis Conditions. *J. Therm. Anal. Calorim.* **2003**, *71*, 707–714. [CrossRef]
30. West, S.L.; White, G.N.; Deng, Y.; McInnes, K.J.; Juo, A.S.R.; Dixon, J.B. Kaolinite, Halloysite, and Iron Oxide Influence on Physical Behavior of Formulated Soils. *Soil Sci. Soc. Am. J.* **2004**, *68*, 1452–1460. [CrossRef]
31. Dayal, A.M.; Varma, A.K. Exploration Technique. *Shale Gas Explor. Environ. Econ. Impacts* **2017**, *2017*, 65–93.
32. Eyssautier-Chuine, S.; Vaillant-Gaveau, N.; Charpentier, E.; Reffuveille, F. Comparison of Biofilm Development on Three Building and Restoration Stones Used in French Monuments. *Int. Biodeterior. Biodegrad.* **2021**, *165*, 1–10. [CrossRef]
33. Fernandes, P. Applied Microbiology and Biotechnology in the Conservation of Stone Cultural Heritage Materials, Mini Review. *Appl. Microbiol. Biotechnol.* **2006**, *73*, 291–296. [CrossRef]
34. Moussa, A.; Badawy, M.; Saber, N. Chromatic alteration of Egyptian Blue and Egyptian Green Pigments in Pharaonic Late Period Tempera Murals. *Sci. Cult.* **2021**, *7*, 1–15.

35. Kiel, G.; Gaylarde, C.C. Diversity of Salt-Tolerant Culturable Aerobic Microorganisms on Historic Buildings in Southern Brazil. *World J. Microbiol. Biotechnol.* **2007**, *23*, 363–366. [CrossRef]
36. Gettens, R.J.; Pease, M.; Stout, G.I. The Problem of Mold Growth in Paintings. *Tech. Stud. Fine Arts* **1941**, *9*, 127–143.
37. Chaurasia, L.; Verma, R.K.; Bisht, V. Microbial Carbonate Precipitation by Urease Producing Bacteria in Cementitious Materials. *Int. J. Adv. Biotechnol. Res.* **2014**, *5*, 671–679.
38. Warscheid, T.H.; Braams, J. Biodeterioration of Stone: A review. *Int. Biodeterior. Biodegrad.* **2000**, *46*, 343–368. [CrossRef]
39. Bai, H.; Liu, D.; Zheng, W.; Ma, L.; Yang, S.; Cao, J.; Lu, X.; Wang, H.; Mehta, N. Microbially-Induced Calcium Carbonate Precipitation by a Halophilic Ureolytic Bacterium and its Potential for Remediation of Heavy Metal-Contaminated Saline Environments. *Int. Biodeterior. Biodegrad.* **2021**, *165*, 20–29. [CrossRef]
40. Rolleke, S.; Muyzer, G.; Wawer, C.; Wanner, G.; Lubitz, W. Identification of Bacteria in a Biodegraded Wall Painting by Denaturing Gradient Gel Electrophoresis of PCR-Amplified Gene Fragments Coding for 16S rDNA. *Appl. Environ. Microbiol.* **1996**, *62*, 2059–2065. [CrossRef]
41. Favero-Longo, S.E.; Matteucci, E.; Pinna, D.; Ruggiero, M.G.; Riminesi, C. Efficacy of the Environmentally Sustainable Microwave Heating Compared to Biocide Applications in the Devitalization of Phototrophic Communities Colonizing Rock Engravings of Valle Camonica, UNESCO World Heritage Site, Italy. *Int. Biodeterior. Biodegrad.* **2021**, *165*, 10–20. [CrossRef]

heritage

MDPI

Article

On Borders and Expansion: Egyptian Imperialism in the Levant during the Ramesside Period

Georgia Xekalaki

Archaeology & Arts, 10 Karytsi Squ, 10561 Athens, Greece; zxekalaki@arxaiologia.gr

Abstract: This paper aims to define the way Egyptians perceived the boundaries of their land and reassesses the impact of Egyptian colonialism during the Ramesside period (c. 1292–1069 BCE). During this era, expansive wars, diplomatic action and land administration/governance reforms led Egypt to control a large part of modern Israel, Palestine, Lebanon and Syria. To refer to this period, historians often use the terms Egyptian "empire" and Egyptian "imperialism", extending terminology coined in the 19th century to describe modern cases of political dominance to Late Bronze Age Egypt. Furthermore, traditional scholarship also presents Egypt's borders in such a way that Egypt appears as a solid territory with fixed borders, despite evidence pointing to a different model of geographical division. Seeking to explore whether the use of modern terms on ancient Egypt may be an anachronism, this paper reviews the scholarship on (a) Egyptian records documenting conquests and (b) contextual archaeological evidence from the southern Near East itself. This review highlights differences between modern and ancient conceptions of land domination. Finally, Egyptian border-related terms are used in a strictly local symbolic cultural context but not in the one of international diplomacy. As for Egypt's boundary, it was mostly formed as a buffer zone rather than a borderline.

Keywords: archaeology; Egypt; New Kingdom; Ramesses; imperialism; border; frontier; boundary; Kadesh; Palestine; Syria; Israel; Mediterranean; history; heritage; culture

Citation: Xekalaki, G. On Borders and Expansion: Egyptian Imperialism in the Levant during the Ramesside Period. *Heritage* **2021**, *4*, 3938–3948. https://doi.org/10.3390/heritage4040216

Academic Editor: Francesco Soldovieri

Received: 8 September 2021
Accepted: 20 October 2021
Published: 25 October 2021

1. Introduction

In the history of academic research, scholarly interest in a local perspective on an imperialistic or colonial situation is a prevalent, but relatively new, area. Following World War II, there was a shift in the study of historical empires and their subjects as a result of geopolitical changes.

The term "imperialism" describes models of domination of a national entity over others that combine territorial expansion, political control, economic exploitation, and cultural influence. As an idea, it stems within the sphere of the 19th c. political thought, to describe modern cases of political dominance, reflecting, nevertheless, the classical paradigm of Rome [1] (pp. 2–3). The term has then been adapted and established to identify domination policies of other ancient territorial powers classified as 'empires', including pharaonic Egypt. Indeed, Egypt's policies towards its southern as well as northeastern neighbors have been persuasively identified as "imperialism" by modern scholarship [2]. However, there is still room to challenge the wide application of the term [3] (pp. ii, 6, 16–30). While "an imperialist power must have imperialist aims and motives" [4], these aims and motives still vary according to case. Since imperial systems are dynamic in nature [5] (p. 8), there is enough space for exploring the exact dynamics of Egyptian expansion during the Ramesside period, from the aspect of aims and motives that also characterize the cultural modality.

This paper seeks to reconsider the way Ramesside Egypt perceived the issue of borders and territorial expansion in relation to the way modern scholarship has presented it within its understanding of Egyptian territorial control and expansion policies as "imperialism".

It aims at highlighting the way Egyptians perceived their borders from c. 1292 to 1069 BCE, and to explore whether using the modern term "imperialism" to describe ancient Egyptian expansion policies is actually an anachronism. Thus, we review the scholarship on (a) Egyptian records documenting conquests and (b) contextual archaeological evidence from the southern Near East itself, recording Egyptian heritage elements. Finally, Egyptian border-related terms are used in a strictly local symbolic cultural context but not in the one of international diplomacy. This review highlights differences between modern and ancient conceptions of land domination and deepens into the ancient cultural diplomacy.

Historical Context

The issue of territorial borders is critical to beginning such an exploration as such borders set a temporary limit of territorial domination and set the controlled area apart from the uncontrolled/hostile one. In the traditional understanding of history, borders' setting, shifting and guarding is often presented as both the aim and the outcome of transformational events. The importance of border-setting for both European nation-states and postcolonial states [6] (pp. 177–178) has led to a common understanding that setting solid territorial borders is key to securing international prosperity. Additional to this assumption is the wide textual and contextual record from pre-modern times (Antiquity and the Middle Ages), such as peace treaties and monumental strings of fortifications perceived as imperial frontier walls (e.g., the Great Wall of China, Hadrian's Wall, UK, the Justinian Wall-Zenobia, Syria) [7] (pp. 233–236), [8] (p. 1).

Regarding ancient Egypt, the textual attestation of terms designating edifices of defensive character separating Egypt from the outside world as well as through an extensive record of written and archaeological heritage evidence indicating a systematic strict control of people and goods entering and leaving the territory of Egypt suggest the existence of strict, solid borders. However, examples from modern challenged territories show that borders can be designated in a complicated way, encircling areas under various levels of control and eternally subject to change (e.g., [6] (pp. 180–181), [9] (pp. 119–125)). Thus, evidence might not be enough to testify to the complete nature of Egypt's and other ancient states' territorial sovereignty.

For about four centuries (from the mid-16th to the 12th century BCE) the Egyptian state was characterized by territorial expansion and domination policies that have been interpreted as imperialistic. The first Pharaohs of the 18th Dynasty, especially Thutmose III, pursued a series of campaigns resulting in several well-reported conquests in Nubia and the Levant [10]. This was followed by a military and diplomatic struggle for maintaining these conquests that lasted from the mid-18th Dynasty to the start of the 20th Dynasty, with breaks during the Amarna period and the last years of the 19th Dynasty. The Ramesside period (19th and 20th Dynasty, 1295–1069 BCE), which is essentially characterized by the struggle to regain what was lost during the Amarna period, is characterized by the stabilization of power with the Kadesh treaty (1269 BCE) and ends with Egypt's last conquests in Palestine lost and settled by new inhabitants [11] (pp. 272–309).

While the 18th Dynasty Pharaohs had an interest in expansion to both the North and the South, most of the Ramessides had a particular interest in maintaining their Levantine conquests. These were cities spreading along two main routes connecting Egypt to Asia Minor (one inland and one along the Syro-Palestinian coast), with major strategic and commercial prominence, yet challenged by nomadic tribes [12] (p. 193), [13] (p. 6). The issue inconsideration, however, is the actual geographical definition of the territory, which Egypt sought to control during the whole phase of the Empire, with a particular focus on the aspirations of the Ramesside pharaohs. Thus, tangible and intangible cultural elements, concerning the Egyptian border-related terms are used in a strictly local symbolic cultural context but not in the one of international diplomacy.

How did the Egyptians perceive the borders of their country during the Ramesside period? What did dominance over a territory mean for Egypt in the 19th and 20th Dynasty? To what extent is the Egyptian territorial dominance model different from the modern concept of imperialism by which is often explained? As this subject is only slightly explored, with a limited bibliography, our work is a step forward in the knowledge of heritage studies from the point of sovereign supremacy in ancient Egypt and Egyptology.

Seeking to define the territory the 19th and 20th Dynasty Pharaohs sought to control, I will attempt a scholarly review of evidence in relation to the idea of borders as presented in recent scholarship, followed by a synthesis of available information. This evidence is both textual and contextual. Textual evidence comes in the form of terms, place names, expressions and text excerpts surviving in the Egyptian linguistic corpus and connected to ideas such as borders, frontiers, territorial expansion and even imperialism. Contextual evidence comes in the form of edifices interpreted as defining frontiers and ranking from simple stelae to larger structures such as forts and walls.

2. Material Evidence and Method

2.1. Egyptian Terms Related to Frontiers/Borders, Territorial Control and Imperialism

As the Egyptian linguistic corpus does preserve words and signs designating home territory, several terms have been linked to defining "border". In general, scholarship has interpreted the word djer (Dr) as referring to an eternal and universal limit and tash (tAS) as a signifier of actual geographic frontier set by deities, but also people [14], and is thus more akin to what is of interest here.

Stephen Quirke [15] (p. 261) interprets tAS as "frontier", also making a clear distinction between frontier and border. According to Quirke, a border is more or less fixed while a frontier is a loose "indefinite zone" "when the polity comes to an end". Later, Galán [2] (p. 9) focuses on evidence from the Empire's formation during the 18th Dynasty and argues that tAS refers to "the area limits of someone's authority or sphere of action". He also highlights that tAS could "be set" (iritAS) and "expand" (s.wsHtAS), with the second having arguably the closest Egyptian parallel to the term "imperialism" (referencing Bleiberg [16]). However, he refrains from discriminating between what in English is called "border" and "frontier". Only recently, Langer and Götz [17] (p. 34) have interpreted the distinction between border and frontier as a continuum rather than a dichotomy between two types of boundaries. In this study, in which previous bibliography is assessed, the following is demonstrated:

- Egypt's tAS is equal to a "sphere of influence", the territory directly or indirectly controlled by the state, while Dr is a "sphere of chaos" beyond the Egyptian control.
- tAS is connected to the cosmological idea of mAat and Dris equally connected to the idea of isft, mAat's opposite and
- The two are fluid and while the Egyptians expand the tAS, the Dr is reduced. Since movement was continuous and Egyptian control fluid, the result was the creation of culturally mixed borderlands.

The assessments above are based on records dating prior to the Ramesside period. Still, key Ramesside examples from royal temple inscriptions seem to confirm:

- There is a set tAS that might be threatened or/and plundered by enemies; e.g., Pharaoh Merenptah accusing the "Nine Bows" of plundering Egypt's borders present an example of the term's use [18] (p. 4).
- The tAS is fluid, and ideally ever-expanding: The word is used in the standard royal phrase defining the Pharaoh as one 'who establishes his boundary as far as he wants, in any land' [irtAS = f rdd(t) Hr = f]. e.g., the "rhetorical" text of Seti I in Karnak, [19] (p. 7). The Kings are also praised for having extended it; e.g., Ramesses II is "extending their borders forever" (sHD.n = k tAS = sn) [20] (p. 197).

On the contrary, the Kadesh treaty, the "international" document signed between the Egyptians and the Hittites, does not employ the term tAS for borders. Instead, the phrase used in the hieroglyphic text of the treaty, as it survives in Karnak, is "trespass the land" (thi r pAtA) for indicating violating the border of Egypt (kmt)and of Hatti (Hti), respectively [20] (p. 227, §13–14)

2.2. Lists of Placenames

Since the 18th Dynasty, lists of foreign place names (often associated with the iconography of bound captives) have been inscribed on temple walls, statue bases or stelae set within the premises of temples or other royal edifices; indeed, this custom continues throughout the Ramesside period, with Seti I and Ramesses II, Merenptah and Ramesses III commissioning such lists on their monuments. These lists, recording place names of the Near East (Northern lists) as well as Nubia (Southern lists) have been recorded, documented [18–20] and also assessed by K. Kitchen who has classified them according to context and content [21].

According to this analysis, context-wise, these place names, often linked with triumph scenes, record the willingness of the pharaohs to report the gods on their success in imposing mAat to a certain geographical sphere outside Egypt. In the case of Seti I, Ramesses II, Merenptah and Ramesses III such lists may also occur in battle reliefs [21] (p. 130), an event further highlighting the link between war and the domination over certain lands. Content-wise, the Northern place names of the Ramesside period reproduce original lists of Thutmose III. In the case of Seti I and Ramesses II, however, lists include additional, original place names [21] (p. 136), in an attempt to reflect contemporary events.

2.3. Egyptian Terms Related to Fortified Frontiers and Areas of Control

Since the Middle Kingdom, the Egyptian record preserves the term inbw-HqA, translated as "The Walls of the Ruler" and connected with a series of fortresses along the Eastern Delta that defined the frontier line with Palestine [14] (p. 318), [22], [23] (p. 159). During the New Kingdom, there is extended evidence referring to a similar defense system known as "The Ways of Horus" (wAtHr). The term is already recorded in the Old Kingdom's Pyramid texts, but only in the New Kingdom is it associated with a heavily fortified gateway to the Northeast, in use as such already in the reign of Thutmose III [24] (p. 7). Alan Gardiner identified and documented Ramesside epigraphic evidence on the Ways of Horus in a series of reliefs commissioned by Seti I on the outer North wall of the Hypostyle Hall of the Temple of Amun at Karnak [25] (pp. 99–116). Gardiner also identified several place names mentioned in the reliefs to others, known from Papyrus Anastasi I [25] (p. 103). Later scholarship confirmed that the Ways of Horus indeed had the form of a series of settlements extending from modern-day Qantara in Ismailia to Rafah in the modern border of Egypt with Palestine (the Gaza Strip) [26] (p. 33), actually ascribing an Egyptian-controlled route, connecting rather than separating Egypt from the outside world. The easternmost site attested in both documents of Pap. Anastasi I and Seti I' s reliefs at Karnak, and solely associated with the end of the route at least during Seti I' s time, is Rafah. The later record bears one more entry, Gaza itself, mentioned as the last part of the route in Pap. Anastasi I [25] (p. 113). Central to that roadway was the Tjaru/Sile, identified with Tell el Heboua, near Qantara.

Tjaru is designated as anxtm, a term derived from the word "lock" or "seal" and generally meaning sites "locking" the access to an area worth controlling. Another toponym associated with anxtm on the Ways of Horus, which is mentioned in Pap. Anastasi I is hTyn. Gardiner identifies it with Megiddo [25] (p. 113) while Morris [27] (p. 804) sees it as identical to the Migdol of Menmaatre mentioned in Seti I's Battle reliefs [19] (p. 10). Beyond the Ways of Horus, however, Morris has recorded five other xtm fortresses functioning during the Ramesside period: "the xtm of the sea" set on the mouth of the Pelusiac branch of the Nile, Tjeku (Tell el-Retaba) on the WadiTumilat transit corridor, thextm at the highland of Coptos on the Wadi Hammamat, and two at Elephantine and Biga on the first Cataract. The above evidence shows that the term was applied to forts controlling the main passages to Egypt, where the documentation of people and goods entering and leaving the country was first taking place [27] (pp. 804–814). Moreover, these Ramesside-era xtm fortresses appear as literally encircling the Nile Valley's Egyptian part and controlling routes at the same time.In that sense, they designate a borderline defensive network combination shaping a heavily controlled zone around Egypt.

It seems, however, that this Egyptian control network did not end with these lines of forts. According to Gnirs [26], a network of vassal cities beyond the Ways of Horus could function in a similar way to the fortresses, stationing Egyptian representatives who would send dispatches to the Pharaonic Residence, or securing military expeditions with provisions. Garrison towns designated asdmiw were founded, however, wherever the network was in need of a "bridge" between two of its towns, which would be subsequent but at a distance from each other [26] (p. 35). This system was forming a buffer zone between the heavily controlled area around Egypt and the outside world.

2.4. Boundary Stelae'

The term "boundary stelae" refers to a type of monumental inscriptions often accompanied with iconography, occurring either on free-standing blocks of stone or carved on rocks and interpreted as marking the end of a territory under control, whether this is Egypt as a country or the different names within it. Such a stela, commissioned by Thutmose III "to mark Egypt's northern border at his time" (carved alongside another one, made by Thutmose I), is known from six textual sources-five royal and one private [28] (p. 337), but not as a physical object. While no such stela has ever been found in Egypt's northern putative border, a number of monuments that marked Egypt's southern border survive in the area of Nubia; these monuments date from the Middle Kingdom up to the 18th Dynasty, but no later. A stela of Thutmose III surviving as a rock cut inscription at Kurgus, Nubia, and referring to his northern and southern "boundary"(tAS) may be seen as a parallel of Thutmose III's lost monument in the Northern Levant [29] (p. 51, Figure 7, p. 52). Kurgus has also given two more stelae set side by side, the first by Thutmose I and the second by Thutmose III. Inscribed with a curse-on-threat formula against the Nubians rather than a border-setting text, these stelae are placed in a mode reminiscent of the Levantine boundary stelae's description and may thus be considered as another parallel. A cartouche of Ramesses II carved next to them shows an attempt of the latter to "endorse" the stelae's message without erecting another [29] (p. 49, Figure 4, p. 50).

The Ramesside kings produced a number of monumental stelae erected in the Levant [30] (pp. 1–3); these stelae are worth reviewing as we seek to identify possible "descendants" of the earlier border stelae known from Nubia. Seti I commissioned the northernmost known such stela, found at Kadesh (Tell Nebi Mend) in Syria [19] (p. 25 § 9), [30] (p. 2) as well as stelae in Tell el Shihab, Syria [19] (p. 17 § 5), Tyre in Lebanon [30] (p. 117) and two more at the Beth Shan in Palestine [19] (pp. 11, 12–15). Ramesses II boasts a larger corpus;like his father, he commissioned a stela in Palestine's Beth Shan, and in Lebanon's Tyre (two fragments inscribed with his cartouche [30] (p. 2, footnote 8) and Byblos (a site which has also given a stela by Thutmose III) [20] (p. 63), [31] (p. 224). In Lebanon, Ramesses commissioned a stela at Adhlun [20] (p. 223) and three rock cut stelae at Nahr el Kalb [20] (pp. 1, 149), [27] (pp. 360–361), [31] (p. 385), the earliest ones in an impressive group of 22 stelae inscribed there by different conquerors across the millennia [32]. Ramesses II is also attributed to a Syrian stela, from Sheikh Said which is known as "the stone of Job", (KRI II, 223). Then, a stela from at-Turra is the only known stela of Ramesses II ever found in what is now Jordan [30].

As demonstrated above, the term tAS is found in Seti I's Tyre [19] (p. 117) and the second Beth Shan stela [19] (p. 16). From the stelae of Ramesses II, the ones from Sheikh Said and Adhlun could be seen as signifiers of land dominance since they bear the iconographical themes of the King offering mAat to a deity and the King slaying enemies before Amun, respectively. Regarding textual evidence, only the stela of Byblos bears an extended text, recording Ramesses II's expedition in the area in his Regnal Year 4 [27] (p. 360). In the rest of the stelae, the text is either limited or not well preserved.

2.5. Archaeological Evidence Related to Fortified Frontiers and Areas of Control

The evidence presented here is coming exclusively from the seminal work of Ellen Morris on Egyptian imperialism [27], as she was able to relate toponyms referring to

Egyptian border control posts with actual sites in the area of the Sinai, Palestine and the Northern Levant. Of these sites, designated as xtmw or dmiw, the only fully researched are Tell el Retabah and Tell el Heboua I [27] (p. 807).

Tell el Retabah, located along the Wadi Tumilat, bears the remains of a fortress identified as the border fortress of Tjeku [27] (p. 396, Figure 28). The area gained prominence during the Hyksos period, and most architectural remains may go back to the 18th Dynasty; however, the earliest recorded portable finds come from the 19th Dynasty. Nevertheless, the settlement was surrounded by an enclosure wall and featured a 19th Dynasty state-sponsored temple dedicated to Atum of Tjeku and a monumental building made of bricks known as the "Great House". Further buildings do not survive, although there are early testimonies of "numerous remains of brick houses", anda necropolis of mudbrick tombs survives 400 m. north of the town [27] (pp. 504–508).

Tell el Heboua Ihas been identified as the ancient Egyptian Tjaru; also a former Hyksos power-base, it is an extensive settlement with a considerable amount of architectural remains (enclosure, granaries) dating before the 19th Dynasty. Research has so far shown that Seti I reconstructed and enhanced the settlement's mudbrick enclosure, while there is considerable architectural evidence to support the presence of a monumental building equivalent to Tell el Retabah's "Great House", as well as further administrative buildings, magazines, storage facilities but also houses. Around the site, research has located a number of smaller settlements identified as satellite communities [27] (pp. 509–511).

Elements common to both sites allowed for the identification of the characteristics of anxtm site. Such characteristics are the following:

- the location in an entry-site to the Nile Valley;
- the presence of a central monumental building serving administrative purposes
- fortification.

Thus, it is possible to identify several archaeological sites as xtm, sharing the characteristics described above, despite the fact that they are not documented as xtm in written sources; an example is Kom el-Qulzoum, a massively fortified site of 20th Dynasty date [27] (p. 742).

Alternatively, other sites known from the archaeological record of the southern Levant might be identified otherwise. A site is identified as a dmi if the central building known as the xtmw is central to an administrative-storage complex which often includes a temple as well as domestic space [27] (p. 816). Small, rectangular fortified compounds with "numerous small administrative, domestic and industrial buildings" surrounding an empty central area are identified as Mkdr (migdol) [27] (p. 819).

Morris evaluates the extensive Ramesside architectural finds, which come from a number of sites spread along the main routes, joining Egypt to Palestine and Syria via the Sinai. At the same time, she assesses portable finds representative of the character of the material culture of these sites. Such sites, where Egyptian heritage is evident, include Deir el-Balah, Tell el-Ajjul, Haruba site A-345, Bir el-Abd [27] (p. 222), Tell el-Farah, Tel Sera, and Tell el-Hesi, (p. 394), Ashdod, Tel Mor, Gezer, Aphek, Jaffa, and Beth Shan, with some in use also during the 18th Dynasty.

Morris finds that, while in the late 18th Dynasty, Egyptians invested in the security of their immediate border territory by supporting bases clustered around the Ways of Horus [27] (p. 270); during the Ramesside period, they set military bases further ahead Gaza (the northern limit of the Ways of Horus), across the southern Levant's maritime way (Via Maris) [27] (p. 397 Figure 29, 717, Figure 54) probably in an attempt to monitor sea routes. Such bases flourish during the 19th Dynasty but fade by the mid-20th, especially after Ramesses III [27] (p. 738). Like the forts along the Ways of Horus, these bases seem to be set a day's journey apart from each other; still, while the Ways of Horus' bases are fortified, the ones further away are not [27] (pp. 387–388).Mostly designated as dmiw, these bases supported a considerable population with mixed (Egyptian and Levantine) cultural traits, living under Egyptian regional command [27] (pp. 826–827). These cultural traits, demonstrated in finds-ranking from locally produced yet stylistically Egyptian pottery and

Egyptianizing objects to large Egyptian-style temples [27] (pp. 713–714), suggest that these settlements' inhabitantshad developed a mixed heritage background [33] (p. 280); within this background, Egyptian influence is strong, yet not longstanding [34] (p. 249), [35] (pp. 78–79, 93, 99). Nevertheless, dmiw formed a buffer zone between the locks of the Nile Valley and "unconquered" lands.

3. Discussion

In search of Ramesside Egypt's "imperialistic aims", it is essential to define how Egyptians perceived the ideal border of their country and whether it was possible to reach it.

The so-called "boundary stelae" do not depict borders in a literal sense (contra [35] (p. 29)). Set either within spaces already designated as sacred or on sites with advanced symbolic importance, they mostly seem to express the willingness of the Pharaoh to declare his presence and dominion to a strategically important area rather than his ability to trespass into foreign territory, let alone declare a universally accepted limit between two territories. In that sense, the term itself might be problematic for stelae erected or located in the Egyptian/Southwest Asian borderland; consequently, stelae by Ramesside pharaohs in the Levant should not be considered as such.

Ramesside topographical lists may express actual contemporary Egyptian ideas on expansion. A comparison of 19th Dynasty material with earlier and later lists context-wise and content-wise show a certain originality that elaborated on the main purpose of such lists: to record the pharaonic ideal purpose of extending power and control as far as possible. Early lists may have recorded sites actually reached by the Egyptians (e.g., the great lists of Thutmose III). In other cases, however, lists record places linked with through nothing more than positive diplomatic relations and/or trade (e.g., the "Aegean list" of Amenhotep III [36] (pp. 313–317), [37,38]). Ramesside lists, especially the ones of the 19th Dynasty, continue serving a traditional symbolic role within the royal context of the Egyptian cultural sphere. Still, they add to our knowledge of new places of interest exclusive in the Ramesside period, which, alongside the known ones, comprise a network of sites linked with each other, where the Egyptians did express control during the 19th and 20th Dynasties.

References to borders may differ between official texts commissioned strictly within Egypt and others compiled within an international environment to serve diplomatic goals. The standard Ramesside royal epithet "the one who establishes his boundary as far as he wants, in any land", used in royal inscriptions, does express the idea of a border. On the contrary, in the Kadesh treaty's text, the phrase "trespass the land of Egypt ... trespass the Land of Hatti" shows the perception of a fixed border, but lacks any particular wording to express it. Instead, there is the idea of two neighboring territories whose limits are well understood and mutually agreed. Geographical definitions to these borders/limits (absent from the Kadesh treaty) had probably been covered by previous treaties of regional scale, established through the two Great Kings' vassals [38] (p. 16). Overall, the idea of the border as a line between Egypt and abroad was largely symbolic, important only so much as it served Egyptian royal ideology, while territory was what mattered in the "real world" of international diplomacy.

According to the Egyptian worldview, the Pharaoh inherits "everything gods have created"at his coronation, and his purpose is to keep it in balance. The world itself is considered harmonious and the property of the Pharaoh [38] (p. 10) at a distance equal to the horizon or the four pillars sustaining the sky [39] (pp. 51–52). However, there was that part of the world outside Egypt, which was considered as ruled by chaos; only inside Egypt could one find balance (mAat). As a result, one of the duties of the Pharaoh was to expand the territory of mAat as much as he could, essentially by expanding the borders of his country.

The pharaonic duty of territorial expansion took on the dimensions ofa ritual drama concerning the fight of good against evil, depicted in temples and other monuments [39]

(p. 8); to express it, the Egyptians used the narrative of real conflicts, whatever the actual outcome of these fights was for the Egyptians. A clear example is Ramesses II's activity at the battle of Kadesh against the Hittites. There Ramesses is presented fighting to expand his borders against an enemy that behaved badly according to the rules of war [39] (p. 162) and he is finally saved by Amun, to save, as a god himself, his army at the end [40] (pp. 44–46). Gods stand with the Egyptians in every attempt for expansion; indeed, even the foreign gods appear to do so, as manifestations of the Egyptian gods [38] (pp. 11–14).

The reality, however, was different. Egypt had a natural border marked by the desert and the sea [39] (p. 86), [14] (p. 318);outside of these natural borders, communities with advanced standards of living, as well as centuries-old networks of cultural contact and political interaction, were forming a world equally balanced yet more dynamic than the one of Egypt. As a result, the will (and struggle) for expansion derived from the fight of harmony over chaos that tradition implied had to be put to a halt. To face political reality, Egypt had to change its expansion ideology from a centralistic to a pluralistic one: Egypt's borders would not have to be moving outwards; instead, the pharaohs would have to control their borders' fluctuation according to the balance of forces between Egypt and its neighboring states [41] (p. 46).

Diplomatic correspondences between Egypt and its neighboring states show that there was a shift to a pluralist approach concerning borders, at least since the reign of Thutmose IV in the 18th Dynasty. The Amarna Letters, the surviving corpus of the 18th Dynasty diplomatic correspondence, famously show that Egypt was an esteemed member of an "Eastern Mediterranean Great Powers' group" that exercised advanced diplomacy through a series of official letters and gift exchange. The letters reflect Egypt's will to sustain control gained during the end of the 15th and the beginning of the 14th c. BCE (LB IB period) across modern-day Palestine [42] (p. 12) and to maintain a power balance in the Levant together with several Near Eastern states of equal if not more prominent status (Babylon, Mitanni, Assyria and the ascending Hatti).

Ramesside kings had to face another reality: following the loss of the Egyptian prestige during the Amarna and the post-Amarna period, a centralistic approach was most definitely promoted by the military elite who rose to power during the end of the 18th and the beginning of the 19th Dynasty as a justification of Egypt's "renaissance". At the same time, the domination of Hatti as a single Great Power in the Near East made the definition of borders in the Egyptian sphere of influence a top priority. The Kadesh treaty—its Egyptian version being carved in hieroglyphs in the precinct of Amunat Karnak and in Ramesses II's mortuary temple—demonstrates that the Egyptians accepted the end of expansion, not only within an international diplomatic context but also within a local cultic one. They had indeed changed the 'centralistic view' deeply imbued in their culture into a "pluralistic" one.

Several researchers have shown the division of Egypt's Levantine conquests into minor administrative territories; however, such scholarly opinions on the identification of these territories vary [35] (pp. 2–3), [27] (pp. 468–470), [37] (p. 47). This administrative division can be considered an indication that the Egyptians were seeking to exercise their control over a wide area that included settlements and countryside. Still, mapping Levantine cities and sites where Egyptian control is demonstrated, based on monuments found there, shows a pattern in the choice of localities [32] (pp. 6–7), [41] (p. 15), [27] (pp. 398, 716–717). These were cities situated along the two major routes used by the Egyptian army [12] (p. 168) and probably set apart within a specific distance equal to a day's trip from one to the other [27] (p. 532), allowing for marching troops or passing by travelers' supply. It is thus worth exploring whether the Egyptians sought to control specific sites of the Levantine territory or the whole region south of the Hittite area of control.

The idea that the Egyptian control expanded over a network of cities rather than a territory as a whole has been analyzed extensively [39] (p. 89). Since each of those towns seemed to play a different role (economical, administrative or strategic) [12] (p. 207), and they could be taxed more easily, the idea seems plausible. At the same time, large parts of Canaan might have remained a "no man's land", frequented by nomadic groups which were related to the Egyptians ina rather symbiotic way [27] (pp. 488–489). In any case, the Egyptians do not seem to totally control the area, controlling only the territories surrounding settlements. It is also worth stating that cities themselves had no permanent borders [12] (p. 193). Therefore, it is much safer to refer to the Egyptian territory as a network consisting of subsequent settlements and their fluid extended area, rather than a compound conquered land.

4. Conclusions

With regard to the meaning of imperialism for the Egyptians and their consequent activity in the Levant, we have successfully assessed: (a)the scholarship on Egyptian records and (b) the Egyptian heritage elements found within Southern Near Eastern contextual archaeological evidence. Following this review, one can conclude the following:

Imperialism, as it is understood today, is applicable to ancient Egypt, as it connects effectively to the themes of conquest and subjection. Since conquest and subjection of foreign lands by border extension were part of the central Egyptian religious idea of maintaining mAat in the world, imperialism was a crucial part of any Pharaoh's foreign policy, interlocking his secular with his cultic duties. Still, cultural imperialism was present but limited as contextual evidence shows that most Egyptian cultural assets, though influential [43] (p. 126), were eventually appropriated rather than reproduced in the framework of the local Levantine/Canaanite cultural expression, and did not affect Southern Near Eastern heritage in the long term [34] (p. 249).It has also been shown that the local population and its elites were instrumental in shaping historical events in their area all through the Late Bronze Age and the Early Iron Age [44] (p. 125). For all the above, I agree with Anne Killebrew in defining Egyptian imperialism being of an "administrative" or "formal" nature [34] (p. 55). In the Ramesside period, Egypt was threatened initially by the Hittites and later by the completely changed situation in the Western Mediterranean (due to the destruction of Hittites and the Aegean and North Syrian centers during the era of the so-called Sea Peoples). Thus, the Pharaohs' cultic role as maintainers of mAat by expanding the borders of the country became largely symbolic; at first, the Egyptian king had to limit his expansion, after negotiations with his Hittite counterpart, and during these procedures, he had to present himself as equal to his counterpart (Kadesh treaty). During the latest part of the Ramesside period, kings also had to accept their diminished control over the Southwestern Asia after the invasions of the Sea Peoples. Due to weakness, the last Ramessidekings were passive in matters of expansion, although they might have focused on consolidating their control "where Egyptian power was still maintained" [33] (p. 280).

The Egyptians were interested in maintaining control over a fluid sphere of influence rather than a territory in the Levant, which would be around a network of cities over two ancient main routes to Syria. During the Ramesside period, this network consisted of two types of settlements under Egyptian control: (a) Levantine settlements with advanced potential for trade activity and (b) Egyptian garrison towns and way-stations supporting military and administrative activity; the network had the form of a "spectrum", where territorial control was more intense the closer one would get to the Nile Valley. From Gaza and westwards up to Qantara, the network of potentially unfortified administrative centers was replaced by one of heavily fortified "gateways" to Egypt, aiming at controlling the access and exit of people and goods to and from the Nile Valley, respectively [45,46]. This buffer zone would be a protective zone in case of a potential invasion against Egypt [33] (p. 11) but also a safe haven for travelers and merchants.

Funding: This research received no external funding.

Acknowledgments: It is an honor to dedicate this paper to Ioannis Lyritzis on the occasion of his retirement from the University of the Aegean, thanking him for his major support to the discipline of Egyptology in Greece. I would also like to thank Louise Bertini, Yasmin El Shazly and Usama

Mahgoub for helping me with library research and bibliography, as well as Katherine Kelaidis and Anna Foka for proofreading my work and offering comments and suggestions.

Conflicts of Interest: The author declares no conflict of interest.

References

1. Harrison, T. Ancient and Modern Imperialism. *Greece Rome* **2008**, *55*, 1–22. [CrossRef]
2. Galán, J.M. *Victory and Border: Terminology Related to Egyptian Imperialism in the XVIIIth Dynasty*; Gerstenberg Verlag: Hildesheim, Germany, 1995.
3. Duff, C.-A. Ceramic Continuity and Change at Shechem (Tell Balatâh): Assessing the Impact of Egyptian Imperialism in the Central Hill Country. Ph.D. Thesis, OISE University of Toronto, Toronto, ON, Canada, 2010.
4. Garnsey, P.; Whittaker, C. Introduction. In *Imperialism in the Ancient World: The Cambridge University Research Seminar in Ancient History*; Cambridge University Press: Cambridge, UK, 1979; pp. 1–6. [CrossRef]
5. Morris, E. *Ancient Egyptian Imperialism*; Wiley Blackwell: Hoboken, NJ, USA; Chichester, UK, 2018.
6. Okhonmina, S. States without Borders: Westphalia Territoriality under Threat. *J. Soc. Sci.* **2010**, *24*, 177–182. [CrossRef]
7. Lewin, A.S. The New Frontiers of Late Antiquity in the Near East. From Diocletian to Justinian. In Proceedings of the Ninth Workshop of the International Network 'Impact of Empire', Durham, UK, 16–19 April 2009; Hekster, O., Kaizer, T., Eds.; Brill: Leiden, The Netherlands, 2011; pp. 233–263.
8. Bellezza, G. On borders: From ancient to postmodern times. In Proceedings of theJoint Workshop on Borderlands Modelling and Understanding for Global Sustainability 2013, Beijing, China, 5–6 December 2013; The International Archives of the Photogrammetry, Remote Sensing and Spatial Information Sciences: Hannover, Germany, 2013; Volume XL-4/W3.
9. Shlay, A.; Rosen, G. *Jerusalem: The Spatial Politics of a Divided Metropolis*; Polity Press: Cambridge, UK, 2015.
10. Hoffmeier, J.K. The Annals of Thutmose III (2.2A). In *Context of Scripture Online*; Consulted online on 11 May 2021; Hallo, E., Ed.; Brill: Leiden, The Netherlands, 2021. [CrossRef]
11. Van Dijk, J. The Amarna Period and the Later New Kingdom. In *The Oxford History of Ancient Egypt*; Shaw, I., Ed.; Oxford University Press: Oxford, UK, 2000.
12. Redford, D. *Egypt, Canaan and Israel in Ancient Times*; Princeton University Press: Princeton, NJ, USA, 1992.
13. Redford, D. Egypt and Western Asia in the Late New Kingdom: An Overview. In *The Sea Peoples and Their World: A Reassessment*; Oren, E.D., Ed.; The University of Pennsylvania: Philadelphia, PA, USA, 2000; pp. 1–13.
14. Shaw, I. Egypt and the Outside World. In *The Oxford History of Ancient Egypt*; Shaw, I., Ed.; Oxford University Press: Oxford, UK, 2000; pp. 314–329.
15. Quirke, S. Frontier or Border? The Northeast Delta in Middle Kingdom Texts. In Proceedings of the Colloquium The Archaeology, Geography and History of the Egyptian Delta in Pharaonic Times, Wadham College, Oxford, UK, 29–31 August 1988; Cotswold Press: Eynsham, UK, 1989; pp. 261–275.
16. Bleiberg, E. Aspects of Political, Religious and Economic Basis of Ancient Egyptian Imperialism during the New Kingdom. Ph.D. Thesis, University of Toronto, Toronto, ON, Canada, 1984.
17. Langer, C.; Fernández-Götz, M. Boundaries, Borders and Frontiers: Contemporary and Past Perspectives. *eTopoi. J. Anc. Stud.* **2020**, *7*, 33–47. [CrossRef]
18. Kitchen, K.A. *Ramesside Inscriptions. Historical and Biographical, Vol. IV*; Blackwell: Oxford, UK, 1982.
19. Kitchen, K.A. *Ramesside Inscriptions. Historical and Biographical, Vol. I*; Blackwell: Oxford, UK, 1975.
20. Kitchen, K.A. *Ramesside Inscriptions. Historical and Biographical, Vol. II*; Blackwell: Oxford, UK, 1979.
21. Kitchen, K.A. Egyptian New-Kingdom Topographical Lists: An Historical Resource with 'Literary' Histories. In *Causing His Name to Live, Studies in Egyptian Epigraphy and History in Memory of William J. Murnane*; Brill: Leiden, The Netherlands, 2009; pp. 129–136. [CrossRef]
22. Hoffmeier, J.K. 'The Walls of the Ruler' in Egyptian Literature and the Archaeological Record: Investigating Egypt's Eastern Frontier in the Bronze Age. *BASOR* **2006**, *343*, 1–20. [CrossRef]
23. Callender, G. The Middle Kingdom Renaissance (c.2160-2055 BC). In *The Oxford History of Ancient Egypt*; Shaw, I., Ed.; Oxford University Press: Oxford, UK, 2000; pp. 148–183.
24. Vinson, S. Transportation. In *UCLA Encyclopedia of Egyptology*; Wendrich, W., Ed.; UCLA: Los Angeles, CA, USA, 2013; pp. 1–14.
25. Gardiner, A.H. The Ancient Military Road between Egypt and Palestine. *JEA* **1921**, *6*, 99–116.
26. Gnirs, A. Coping with the Army: The Military and the State in the New Kingdom. In *Ancient Egyptian Administration*; Moreno García, J.C., Ed.; Brill: Leiden, The Netherlands, 2013; pp. 639–717.
27. Morris, E. *The Architecture of Imperialism: Military Bases and the Evolution of Foreign Policy in Egypt's New Kingdom*; Brill: Leiden, The Netherlands, 2005.
28. Vogel, C. This far and not a step further! The Ideological Concept of Ancient Egyptian Boundary Stelae. In Proceedings of the Conference at the University of Haifa, Haifa, Israel, 3–7 May 2009; pp. 320–341. [CrossRef]
29. Davies, V. Kurgus 2000: The Egyptian Inscriptions. *Sudan Nubia* **2001**, *5*, 46–59.
30. Wimmer, S.J. A New Stela of Ramesses II in Jordan in the Context of Egyptian Royal Stelae in the Levant. In Proceedings of the Third International Congress on the Archaeology of the Ancient Near East (3 ICAANE), Paris, France, 18 April 2002. [CrossRef]

31. Porter, B.; Moss, R. *Topographical Bibliography of Ancient Egyptian Hieroglyphic Texts, Statues, Reliefs and Paintings, Vol VII: Nubia, the Deserts and Outside Egypt*; Griffith Institute—Ashmolean Museum: Oxford, UK, 1952.
32. Lebanon: Project at Nahr El Kalb, Factum Foundation. Available online: https://www.factumfoundation.org/pag/232/lebanon-project-at-nahr-el-kalb (accessed on 1 August 2021).
33. Martin, M. Egyptian and Egyptianized Pottery in Late Bronze Age Canaan: Typology, Chronology, Ware fabrics and Manufacture techniques. Pots and People? *Agypt. Und Levante/Egypt Levant* **2004**, *14*, 265–284.
34. Killebrew, A. *Biblical Peoples and Ethnicity. An Archaeological Study of Egyptians, Canaanites, Philistines and Early Israel 1300–1100 B.C.E.*; Society of Biblical Literature: Atlanta, GA, USA, 2005.
35. Higginbotham, C. *Egyptianization and Elite Emulation in Bronze Age Palestine*; Brill: Leiden, The Netherlands, 2000.
36. Astour, M. Aegean Place-Names in an Egyptian Inscription. *AJA* **1966**, *70*, 313–317. [CrossRef]
37. Cline, E.; Stannish, S. Sailing the Great Green Sea? Amenhotep III's "Aegean List" from Kom el-Hetan, Once More. *J. Anc. Egypt. Interconnect.* **2011**, *3*, 6–16. [CrossRef]
38. Kemp, B.J. Imperialism and Empire in New Kingdom Egypt (ca. 1572-1087 B.C.). In *Imperialism in the Ancient World. The Cambridge University Research Seminar in Ancient History*; Garnsey, P.D.A., Whittaker, C.R., Eds.; Cambridge University Press: Cambridge, UK, 1978; pp. 7–57.
39. Liverani, M. *Prestige and Interest: International Relations in the Near East ca. 1600-1100 B.C.*; Sargon: Padova, Italy, 1990.
40. Ockinga, B. On the Interpretation of the Kadesh Record. *Chron. D'Egypte* **1987**, *62*, 38–48. [CrossRef]
41. Liverani, M. Moving Borders. In *International Relations in the Ancient Near East, 1600–1100 BC. Studies in Diplomacy*; Palgrave Macmillan: London, UK, 2001. [CrossRef]
42. Weinstein, J.M. The Egyptian Empire in Palestine: A Reassessment. *BASOR* **1981**, *241*, 1–28. [CrossRef]
43. Koch, I. *Colonial Encounters in Southwest Canaan during the Late Bronze Age and the Early Iron Age*; Brill: Leiden, The Netherlands, 2021. [CrossRef]
44. Levy, E. A Note on the Geographical Distribution of New Kingdom Egyptian Inscriptions from the Levant. *JAEI* **2017**, *9*, 14–21.
45. McGovern, P.E.; Fleming, S.J.; Swann, C.P. The Late Bronze Egyptian Garrison at Beth Shan: Glass and Faience Production and Importation in the Late New Kingdom. *BASOR* **1993**, *290*, 1–27. [CrossRef]
46. Forstner-Müller, I. Centre and periphery: Some remarks on the Delta and its borders during the Ramesside Period. In *The Ramesside Period in Egypt: Studies into Cultural ajnd Historical Processes of the 19th and 20th Dynasties*; Kubisch, S., Rummel, U., Eds.; Walter de Gruyter: Berlin, Germany; Lincolnshire, UK, 2018; pp. 103–112.

heritage

Article

The Antikythera Mechanism: The Prove of the Accuracy of the Astronomical Calculations Based on It

Kyriakos Efstathiou [1,*], Marianna Efstathiou [2], Alexandros Basiakoulis [2] and Neofytos Kokkinos [1]

1 Digital Heritage Research Laboratory (Cultural Informatics), Department of Electrical Engineering and Computer Engineering and Informatics, Cyprus University of Technology, Limassol 3036, Cyprus; nc.kokkinos@edu.cut.ac.cy

2 Laboratory for Machine Tools and Manufacturing Engineering, Mechanical Engineering Department, Aristotle University of Thessaloniki, 54641 Thessaloniki, Greece; mefsta@yahoo.gr (M.E.); alexbasiak@yahoo.gr (A.B.)

* Correspondence: kyriakos.efstathiou@cut.ac.cy; Tel.: +357-9656-0960

Abstract: The Antikythera Mechanism is the oldest extant complex geared device, an amazing analogue computer. It was built approximately 2150 years ago. The device was operated manually by a user, setting a date in a dial. All necessary calculations were made using a set of gears (at least 39), while the results were displayed on several scientific scales. The Mechanism was used to calculate astronomical phenomena, such as solar and lunar eclipses. After an extensive description of the Mechanism, the main objective of the following paragraphs is to demonstrate the accuracy of its predictions.

Keywords: Antikythera Mechanism; Gears; Ancient Astronomy; Ancient Technology; Egyptian Calendar

Citation: Efstathiou, K.; Efstathiou, M.; Basiakoulis, A.; Kokkinos, N. The Antikythera Mechanism: The Prove of the Accuracy of the Astronomical Calculations Based on It. *Heritage* **2021**, *4*, 3848–3878. https://doi.org/10.3390/heritage4040211

Academic Editor: Christofilis Maggidis

Received: 6 August 2021
Accepted: 17 October 2021
Published: 21 October 2021

1. Introduction

The Antikythera Mechanism is a gear device, made between 100 and 200 BC, which was operated by hand. A user could put a pointer to the desired date and the Mechanism simultaneously calculated various astronomical phenomena. More than 39 cooperating gears were rotated simultaneously to rotate seven pointers with eight indications on various scientific scales. It was used to calculate the diurnal and annual motions of the Sun, the Moon and probably the planets among the stars. The Mechanism implemented the astronomical knowledge of ancient Greeks about the motion of these heavenly bodies with astonishing accuracy, considering the anomalous orbit of the Moon using a system of eccentric gears. It could also predict eclipses of the Sun and the Moon based on the Saros period, which was found in one of its scales. It calculated the dates of the major crown games that took place in ancient Greece (e.g., Olympic Games). Finally, it was accompanied by an extended User's Manual [1,2].

The first part of the present paper provides an extensive overview of the knowledge about the Mechanism regarding its discovery, investigation, importance and handling. The construction of the Mechanism, as well as the astronomical and technological knowledge that were necessary for its construction are also described. In addition, the front and back sides of the Mechanism are described and information is given concerning the Egyptian, Julian and Gregorian calendars.

The second part of the paper deals with the accuracy of the Mechanism's predictions. In order to verify its accuracy, an application has been developed in a virtual reality environment, which simulates the operation of the physical models. The indicators on both sides of the Mechanism are presented in an image format and not as a numerical result. The results of the forecasts are automatically compared with those of the NASA website The results support the view that the Antikythera Mechanism was a scientific instrument for accurate astronomical predictions.

1.1. The Antikythera Shipwreck and the Discovery of the Mechanism

In first century BC a large Roman ship battled with the waves on the rough sea between the mainland of Greece and Crete. Finally, the boat sank on the shores of the small Greek Island Antikythera. The ship was loaded with works of art and other precious artifacts. Two thousand years later, at the Easter of 1900 AD sponge-divers from the Greek Island Symi, discovered accidentally the ancient shipwreck off the coast of the Greek Island of Antikythera. Underwater excavation began at the end of November 1900, and a few months later, important findings were recovered, such as the famous Antikythera Ephebe (Figure 1), most of which are now exhibited at the National Archaeological Museum of Athens [3]. Based on Pergamon coins that were retrieved, the wreck is dated between 86 and 67 BC [3,4].

Figure 1. 1900 underwater excavation on the shores of the Greek island of Antikythera (**left**). The Antikythera Ephebe (**right**) [3].

Among the findings, a strange bulk of material, broken, worn and calcified, was located with obvious signs of bronze plating (Figure 2) [3]. In the first publication of the Antikythera shipwreck [5], the existence of the Mechanism was mentioned with the suggestion that it was an astronomical instrument. The Antikythera Mechanism, after 2000 years on the seabed, was expected to change the accumulated knowledge so far on the technological skills of the ancient Greeks.

Figure 2. The fragments of the Antikythera Mechanism, National Archaeological Museum of Athens [3].

1.2. The Investigation of the Antikythera Mechanism

Derek de Solla-Price was the first scholar to study the function of the Mechanism extensively, with the assistance of Charalambos Karakalos from the Research Centre Demokritos, Greece. He worked for more than 30 years and eventually published an extensive account, known as "Gears from the Greeks" [6]. He declared that "the Antikythera Mechanism is the oldest proof of scientific technology that survives today and completely changes our view of ancient Greek Technology".

The baton was taken by Michael Wright and Alan Bromley. Unfortunately, Alan Bromley died in 2002. However, Michael Wright published a series of papers, where he correctly postulated that the back dials of the Mechanism were spirals and that the upper dial was built to follow the draconic lunar month. He also elaborated on the pin and slot mechanism (see Section 2.1.2) and proved its epicyclic function [7]. He made strides in reconstructing the Mechanism and he produced superb bronze replicas.

In 2001, John Seiradakis (Aristotle University of Thessaloniki, Greece), Mike Edmunds and Tony Freeth (Cardiff University, UK) and Xenophon Moussas and Yanis Bitsakis (National and Kapodistrian University of Athens, Greece) created the "Antikythera Mechanism Research Group". They received a grant from the Leverhulme Foundation, U.K. and permission to undertake a new investigation from the Ministry of Culture of Greece. After permission was granted, Eleni Magkou and Mary Zafeiropoulou (National Archaeological Museum of Athens, Greece) and Agamemnon Tselikas (Cultural Foundation of the National Bank of Greece) joined the team, which was soon supported by an international team of astronomers, archaeologists, mathematicians, physicists, chemists, computer engineers, mechanical engineers, epigraphologists and papyrologists.

In September 2005, they undertook a major new investigation of the Antikythera Mechanism, using an innovative and state of the art high power micro-focusing X-ray tomography, especially constructed by X-Tek Systems [8] (Figure 3, left) and the Hewlett Packard, USA, PTM Dome technique [9] (Figure 3, right). In November 2006, the results of the investigation were announced during an international conference in Athens and published in the international journal Nature [10]. The three-dimensional images that were obtained when the fragments of the ancient mechanism were examined revealed internal details of gearing and inscriptions that remained hidden on the seabed of the Antikythera more than two thousand years. All inscriptions are written in Greek. A new font (True type font) has since been developed at the Aristotle University of Thessaloniki, Greece, reproducing the fine art letters.

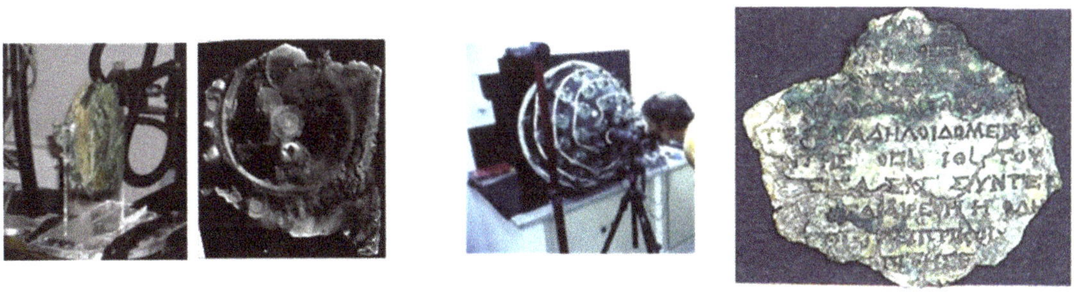

Figure 3. X-ray tomography of the fragment A by Roger Hadland (**left**) and investigation of the fragment 19 using the PTM Dome technique by Tom Malzbender and Dan Gelb (**right**).

Realizing the importance of the Antikythera Mechanism in the Aristotle University of Thessaloniki (AUTH), Greece, an active group of researchers was formed. Several new results have been revealed and published, or currently under publication [11–19].

2. Results

2.1. Description of the Antikythera Mechanism

Based on the results of these investigations [2,3,10,11,13–15], the Mechanism contained a manual with detailed instructions. It had front and back doors with astronomical, geographical and technological inscriptions covering much of the exterior of the Mechanism Its dimensions were approximately 30 × 20 × 10 cm; slightly larger than a current Laptop Similar ancient mechanisms have not yet been found.

The Antikythera Mechanism was a complicated instrument. Therefore, it is not surprising that it was accompanied by an extensive User's Manual. New inscriptions that had not been read for more than 2000 years were revealed, mainly with the X-ray microfocusing tomography. Approximately 3500 letters and symbols have now been deciphered. They all fall into three broad categories: astronomical inscriptions, geographical inscriptions and technical inscriptions. Several astronomical terms have been read referring to the Sun, the Moon, the ecliptic (or Zodiac Cycle), the Metonic and Saros cycles and other astronomical phenomena. The word "ΣΤΗΡΙΓΜΟΣ" (stationary point) is mentioned several times, obviously referring to stationary points of the planetary orbits.

The Mechanism had 7 pointers, which provided 8 indications in its scales (the pointer of the Moon gives two indications). It had three main dials; one at the front side with two concentric scales and two at the back in the form of spirals.

2.1.1. The Front Site, the 12 Zodiac Constellations and the Egyptian Calendar

A very careful analysis of the gears' co-action revealed their use in calculating the exact (within a degree) position of the Sun and the Moon on the Zodiac Cycle. This position was shown by the pointers of the front dials [1–3,10,11,13–17]. Simultaneously, a crown gear was found to drive a black and white colored spherule, showing the current phase of the Moon (see Figure 4).

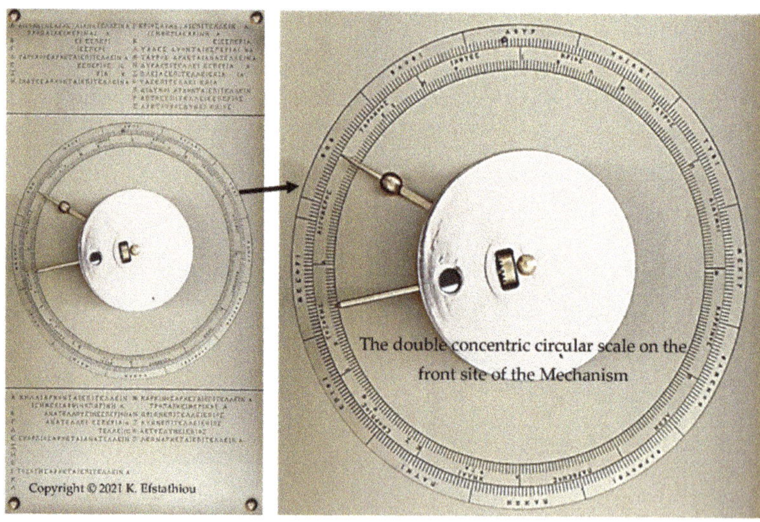

Figure 4. The front side of the Mechanism.

Parts of the front plate of the Mechanism, which testify to its structure, have been saved only in fragment C, with the exception of some very small parts of the 'parapegma' (see Section 2.3.2) that have been read in some of the small fragments. A 'parapegma' is a precursor to the modern-day almanac. Originally, it was a table that related to star phases and corresponding weather. In fragment C, about 1/4 is preserved by a double concentric circular scale that was probably placed approximately in the center of the front plate, as shown in Figures 4 and 5 [16].

Figure 5. PTM image from the C fragment showing a part of the rings on the front side of the Mechanism that has been saved. Some of the letters and subdivisions of the two rings have been imprinted in white to be clearly visible [16].

The date scale had the shape of a ring (Figure 6), which is removable. Behind this date ring, there were 365 holes and the ring was secured at one of the holes using a pin. Every four years, the operator was able to remove the ring and move it by one hole, to take leap years into account. The leap year was established with the Julian calendar by Julius Caesar in 40 BC, a date later than that of the construction of the Mechanism. Ptolemy-III-Euergetes (246–221 BCE) reformed the calendar introducing the leap years." The Canopus decree, a declaration published by a synod of Egyptian priests, suggests that the true duration of the year (365 $^{1}/_{4}$ days) was recognized, and an extra day was added to the calendar every four years. The new calendar failed, which achieved popular acceptance [20]. Nevertheless, the manufacturers of the Mechanism, 100 years later, were aware of the phenomenon.

The internal circular scale depicts the Zodiac Cycle has 360 subdivisions with each of the 12 zodiac constellations having 30 subdivisions. The Zodiac Cycle is the path followed by the Sun during its 'apparent' motion in the sky during a solar year, and is thus essentially the same as the solar year. Namely, the 360 degrees 'apparent' movement of the Sun around the Earth, projected on the Zodiac Cycle, corresponds to a period of about 365 days (solar year).

The part of the Zodiac Cycle that has been saved (Figure 5) is the part of the zodiac constellations Virgo *(ΠΑΡΘΕΝΟΣ)*, Libra *(ΧΗΛΑΙ)*, Scorpio *(ΣΚΟΡΠΙΟΣ)* and Sagittarius *(ΤΟΞΟΤΗΣ)*, with Virgo and Sagittarius only partially surviving.

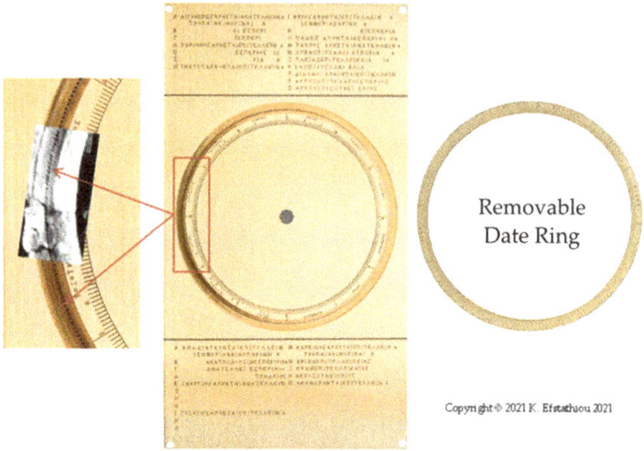

Figure 6. The front site dial plate and the removable date scale ring.

The Egyptian Calendar

As mentioned before the months on the Mechanism are those of the Egyptian calendar, written in the Greek language. The ancient Egyptian calendar [20–26] is important as it included, for the first time in history, the basic principle that the length of the year is equal to 365 days. In addition to this civil calendar, the ancient Egyptians simultaneously used a second calendar based on the phases of the moon.

The ancient Egyptians were the first of the ancient peoples to release their calendar from the synodic lunar month. They associated the course of the year with two almost simultaneous natural phenomena:

- the annual floods of the Nile and
- the appearance of Sirius shortly before sunrise.

The phenomenon of almost simultaneous rises of the Sun and Sirius is called Heliacal rising of Sirius (morning rising, ΕΩΙΑ ΕΠΙΤΟΛΗ or ΕΩΘΙΝΗ ΕΠΙΤΟΛΗ). The Heliacal rising of Sirius occurred around the end of July and determined the beginning of the Egyptian year, which for this reason was called "Dog" or "Cynical", since Sirius (known colloquially as the "Dog Star") is the brightest star in the constellation Canis Major (Big Dog, ΜΕΓΑΣ ΚΥΩΝ), while at the same time it was identified with the dog-headed god Anubis [25].

An ivory plaque found in a tomb of the first dynasty in Abydos greets Sirius as "the bearer of the new year and the flood of the Nile". For the same reason, Plutarch, in his Isidore, characterizes Sirius as an "aqueduct" *(ΥΔΡΑΓΩΓΟΝ),* in order to emphasize that he was to blame for the floods of the Nile [25].

The calendar appearance date of Sirius, however, changed over the years and deviated from the real date, due to the fact that the Egyptians used an approximate solar year of 365 days and not 365.25 days.

The primary Egyptian calendar the older of the two systems was the lunar calendar, divided into 12 synodic months. Each lunar month began with the new moon. Since the lunar calendar was 10 or 11 days shorter than the solar year, a 13th month was intercalated every several years to keep the lunar calendar in rough correspondence with the agricultural seasons and their feasts. The first month of the lunar calendar, as well as of the solar calendar, which was established later, was dedicated to the great divider of time, the god of Astronomy Thoth *(ΘΩΘ),* perhaps for religious reasons.

The Egyptian solar or civil calendar contained 360 days, with twelve equal months of 30 days. However, the dates, as the priests observed, were constantly ahead of the respective climatic seasons, and the annual error of 5 days created a difference of one whole year every 72 solar years ($5 \times 72 = 360$ days). To correct this discrepancy, they introduced the 5 extra days, probably during the time of the first historical pharaoh Menes (2900 BC) and called them "induced" (epagomenal *επαγόμενες*) days.

However, in order to justify them and not to upset the gods, they connected in their mythology these days with the myth of the goddess of the sky, Nut. Nut cheated on her husband Ra, the great god of the Sun, with Thoth from whom she became pregnant. To punish her, Ra ordered that should could not give birth to her child on any month of the year, but the wicked Thoth tricked the Moon into gambling and won her 5 days each year. However, these days did not belong to any month and were outside the calendar, so Ra's order did not apply to them. According to Egyptian mythological tradition, Nut gave birth to her child on the first of these five epagomenic days. The five sacred induced days were dedicated to the birthdays of the five main Egyptian gods, Osiris, Horus, Isis, Seth and Nephthys.

The names of the Egyptian months, mentioned by the great Greek astronomer of antiquity Claudius Ptolemy (second century AD), are listed in Table 1:

Table 1. The names of the months of the Egyptian calendar written in the Greek language.

Egyptian Months	
ΘΩΘ	Thoth
ΦΑΩΦΙ	Phaophi
ΑΘΥΡ	Athyr
ΧΟΪΑΚΙ	Choiak
ΤΥΒΙ	Tybi
ΜΕΧΙΡ	Mechir
ΦΑΜΕΝΩΘ	Phamenoth
ΦΑΡΜΟΥΘΙ	Pharmuthi
ΠΑΧΩΝΣ	Pachons
ΠΑΫΝΙ	Payni
ΕΠΙΦΙ	Epiphi
ΜΕΣΟΡΗ	Mesore
+5 induced *(epagomenal)* days.	

Sirius was called Thoth as his Heliacal rising coincided with, or indicated, Thoth, the first month of the Egyptian year. He was worshiped as Sihor, the star of the Nile, as Sothi, Sothis and Siris. The ancient Egyptians still believed that Sirius was the abode of souls, while at the same time they believed that radiation from the region of Sirius enlivened the creatures of the Earth.

The total duration of the year for the ancient Egyptians was, thus equal to 365 days and the error was reduced to about 0.25 days per year, since the duration of the tropical year was considered equal to 365.25 days.

The strange thing is that although the priests-astronomers were aware of this error of about 6 h per year, they nevertheless did not make any correction. This correction was probably mad due to the fact that the Pharaohs, ascending the throne, swore in the temple of Isis that they would keep exactly the current calendar, not due to the religion of the people.

The Julian Calendar

As mentioned before, the Egyptian civil calendar was altered by Julius Caesar about 46 BC with the addition of a leap-year day occurring once every four years; the revised system forms the basis of the Western calendar still used in modern times. By the 40s BC the Roman calendar was three months ahead of the solar calendar. Caesar, introduced the Egyptian solar calendar, taking the length of the solar year as 365 $^{1}/_{4}$ days. The year was divided into 12 months, all of which had either 30 or 31 days except February, which contained 28 days in common (365 day) years and 29 in every fourth year (a leap year, of 366 days). Leap years repeated February 23; the dated 29 February did not exist in the Julian calendar.

The Gregorian Calendar

The length of the Julian-calendar year was overestimated by 11 min 14 s. This error in mid-1500 had shifted the dates of the seasons by about 10 days from Caesar's time. Pope Gregory XIII' restored in 1582 the calendar to the seasonal dates of 325 AD, an adjustment of 10 days. Nearly all Eastern Orthodox churches use the Julian calendar to establish the dates of movable feasts such as Easter.

2.1.2. The Back Site

At the back side of the Mechanism (Figure 7) were two spiral scales. The upper spiral consists of 5 windings, with its total length divided in 235 sections. Using this dial, the user could read the position of the Moon within the Metonic Cycle (432 BC) of 19 tropical years of 365.2422 days, which is almost equal to 235 synodic (lunar) months of 29.5306 days.

The difference between the two periods (of 19 tropical years and 235 synodic months) is only 2 h. This knowledge allowed the calculation of the exact day of full Moons, a very useful knowledge for agricultural or nautical activities 2000 years ago, when no electricity was available. The accuracy of the position of the Moon was achieved by a pin-and-slot mechanism [7,27] that reconstructed Hipparchus' (190–120 BC) first anomaly of the Moon's motion (due to its elliptical orbit around the Earth). This anomaly is, in fact, Kepler's second law.

Calippus (370–300 BC) 100 years later corrected Metons calendric system. Each of the four periods of Meton, i.e., every 76 years, one day needed to be removed. The Callippic pointer of the subsidiary dial within the upper back spiral of the Antikythera Mechanism is indicated when the correction takes place.

A subsidiary dial within the upper back spiral of the Antikythera Mechanism displayed the celebration date of the ancient Panhellenic crown games. On the circumference of the dial, the words Olympia, Pythia, Isthmia, Nemea, Naa and Halieia were deciphered. Internally, in each quadrant, the four years of the Olympic cycle are indicated (see Section 2.3.2).

The lower back dial is a Saros eclipse-prediction dial, arranged as a four-turn spiral. This dial contained the 223-month eclipse Saros Cycle (of approximately 6585.3213 days, or nearly 18 years and 11 1/3 days). A total of 223 lunar months (one Saros Cycle) after an eclipse, the Sun, Earth, and Moon return to approximately the same relative geometry, and a new, nearly identical, eclipse cycle begins. The Saros Cycle was marked with the dates (month, day, and hour) when a possible solar or lunar eclipse would occur. The markings were engraved with symbols (*"H" ΗΛΙΟΣ* Sun, "Σ" *ΣΕΛΗΝΗ* Moon, etc.).

A subsidiary Exeligmos dial, within the Saros dial, extended the eclipse prediction capabilities to three Saros Cycles, indicating that 8 and 16 h should be added, respectively, in the second and third Saros Cycles to the eclipse times, as indicated by the inscriptions.

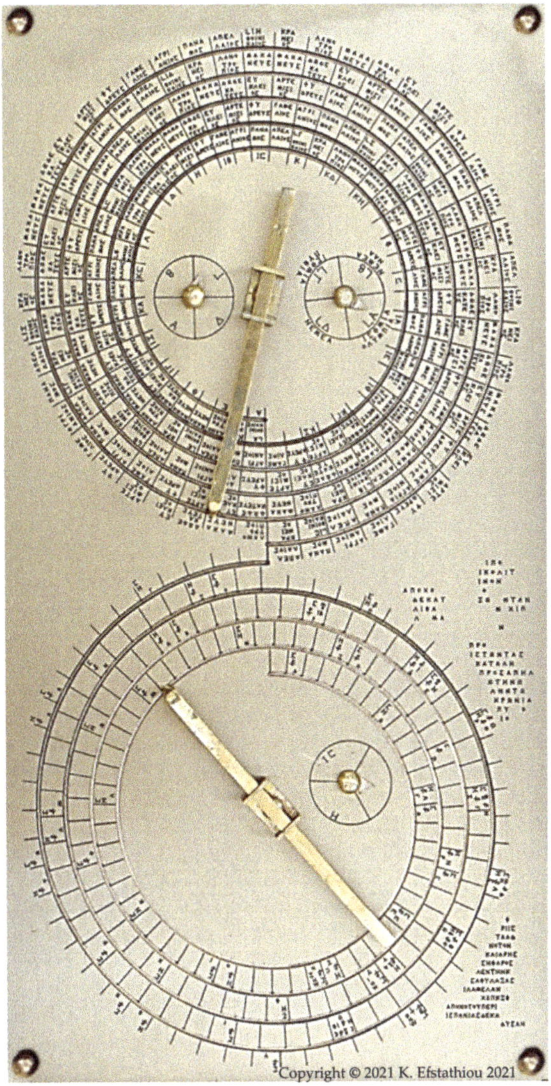

Figure 7. The back side of the Mechanism.

The most impressive part of the Mechanism is related to the moon phases and the moon's movement anomaly. In the 17th century, in relation to the Sun and Moon movements, Johannes Kepler claimed that the "holy circular motion was not circular" and suggested that it was an ellipse orbit (first and second law of Kepler). Inside the Mechanism, a gear system is identified, which simulates this motion with astonishing accuracy.

The display of this movement, considering the moon's movement anomaly caused by its eccentric orbit around the Earth, was achieved by the use of two eccentric gears (Figure 8).

Figure 8. The pin-and-slot mechanism [17].

The axial distance of the rotation's axis of the two gears amounts to approximately 1.1 mm. The lower gear has a pin that engages with a slot on the upper gear, forcing it to rotate by the pin-and-slot arrangement. The epicyclical movement of the upper gear tracked the motion of the Moon in the sky with great accuracy [17,27,28].

The latest outcomes of the AUTH team are related to the investigation of fragment D. In all modern studies of the Mechanism, fragment D is considered a lone fragment and does not exist in any constructed model of the Mechanism. Following a long-term study, it was proven that Fragment D is indeed part of a mechanical arrangement that calculates the equation of time, i.e., the deviation between the real-time shown by the sundials and the average time shown by the current clocks [1,16]. Newer research on fragment D has recently been published [29] and suggests an alternative use of fragment D.

The gears found in the Antikythera Mechanism are the earliest known to resemble the shape and design of modern gears. Their triangular teeth were designed to transmit angular motion, not power. Detailed studies of the fragments of the Mechanism revealed that it had at least 39 gears (37 gears and 2 crown gears), 27 of which have been identified in the largest of the calcified fragments (Fragment A). The existence of ten more gears has been determined using astronomical calculations [10,11,17,30,31]. A functional diagram of the gear trains is presented in Figure 9.

The Antikythera Mechanism also incorporated 2 crown gears, as well as 19 shafts and axles with complex geometry. Two pairs are concentric, which means that one out of the two shafts passes through the other. The two shafts rotate independently. Moreover, there is an axis, which has two eccentric cylindrical bearing points for two gears. At various locations, mainly inside the Mechanism, there were also various components that supported the shafts and other parts of the device to make the construction robust and functional.

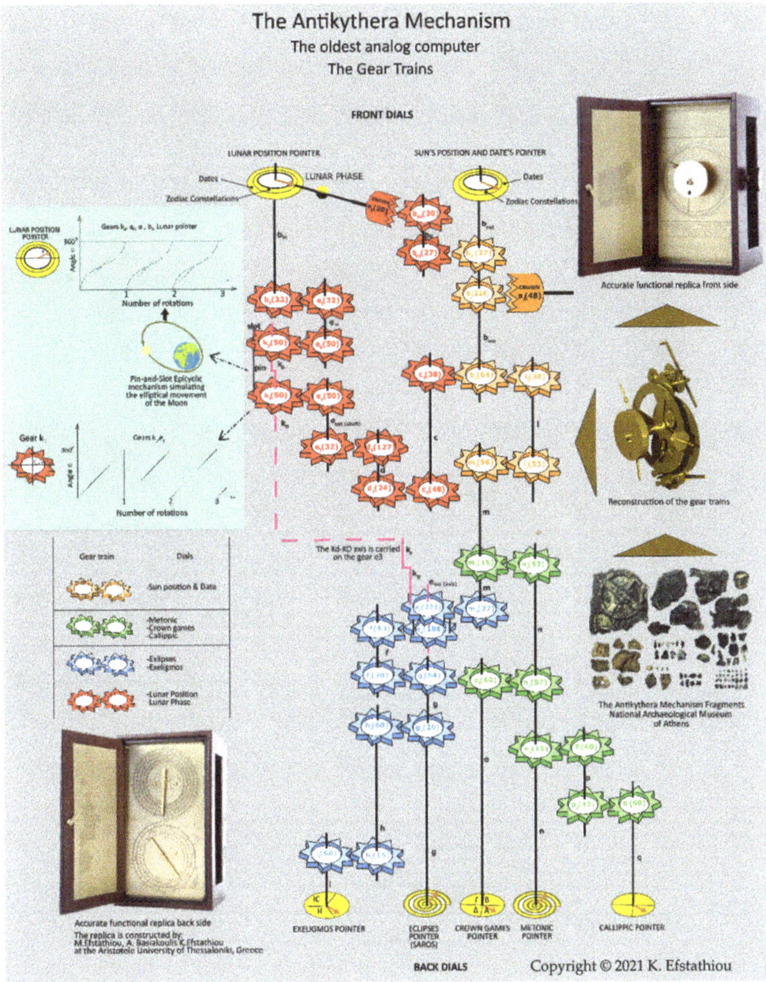

Figure 9. A functional diagram of the gear trains.

2.2. *The Antikythera Mechanism, from Physical to Digital 3D Representations*

Based on the three dimensional images obtained from the X-ray tomography and the PTM Dome techniques, and using the Volume Graphics specific software system VGSTUDIO MAX (https://www.volumegraphics.com/en/products/vgstudio-max.html), the 3D digital objects of the four basic fragments (Figure 2) were developed. The entire Mechanism was designed using the same data [17]. Figure 10 shows the three-dimensional digital reconstructions of the gear trains and the entire Mechanism. An example of how the front of the Mechanism and gear system was designed based on components found in the tomographies, is shown in Figure 11.

Figure 10. Three-dimensional digital reconstruction of the Mechanism [17] (Copyright © 2021 M. Efstathiou).

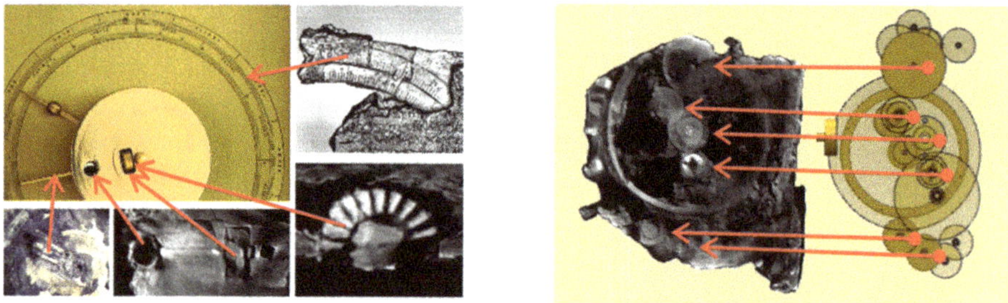

Figure 11. Mechanical parts localized in the tomographies and corresponding designed parts.

2.3. Importance, Purpose, Utility, Operation and Handling of the Mechanism

2.3.1. Importance and Purpose of the Mechanism

The Antikythera Mechanism is an epitome of high technology in antiquity, as the Parthenon is an epitome of Architecture. Considering the theoretical and technological knowledge required for the construction of the Mechanism, it can easily be ranked among the Wonders of Ancient World [1,2].

The Antikythera Mechanism was the first surviving geared analogue computer in history. The next extant geared device is a Byzantine clock calendar, which was built in the fifth or sixth century. More than 800 years later, the next mechanical calculators were built. At the beginning of the 13th century, the astronomical indicator of Wallingford, 50 years later (1348–1364) the astronomical clock of Dondi, and in 1410 the Prague astronomical clock of similar complexity to the Mechanism. Later, in the 17th century, the calculator of Schickard (Kepler's collaborator) and the Pascaline of the great French scientist Pascal were built.

The purpose of the construction of the Mechanism is not yet known. The identity of the manufacturer is also unknown (see Section 2.4), but considering the variety of skills and knowledge required, in order to build such a complex device, a collaboration between an astronomer/mathematician and a mechanic was necessary. The inscription on the back cover is in a different hand to the rest, showing that at least two people worked on the machine [1].

2.3.2. Utility, Operation and Handling

The main utility of the Antikythera Mechanism was to calculate the exact position of the Sun, the Moon and possibly the planets in the sky, as well as the phases of the Moon and the lunar or solar eclipses. It is interesting that besides the predictions of astronomical events, the Mechanism could determine dates related to religious, social and agricultural rituals and events.

The subsidiary dial within the upper back spiral of the Antikythera Mechanism displayed the dates of the Olympic Games, which were held during the first or the second Full Moon after the summer solstice. To learn the exact date, someone had to be able to calculate this phase of the full moon. The Mechanism could accurately calculate this date. Not only the Olympic Games but the crown games of Isthmia (Corinth), Nemea (Nemea), Pythia (Delphi), Naa (Dodona) and Halieia (Rhodes) (Figure 12) were included in the subsidiary dial [17].

Figure 12. The subsidiary dial displayed the dates of the Panhellenic games (**left**). Three runners featured on an Attic black-figured amphora. British Museum, by Marie-Lan Nguyen (2011), CC BY 2.5, https://commons.wikimedia.org/w/index.php?curid=15468713, accessed on 5 August 2021 (**right**).

In Figure 13, the Parapegma of the Mechanism is shown [15]. The Parapegmata were essentially calendars of astronomical and meteorological events, which were widely used in ancient Greece. The astronomical events referred in the Parapegmata are events that associate the rise and set of stars or constellations with the sunrise or sunset. For many years, seasons with different climatic conditions were an important time unit as they played a crucial role in people's lives. Over time, however, it was found that the start date of each season could not be determined with the usage of classic calendars, based on lunar months. Therefore, people turned to stable natural phenomena to define the seasons. Some of these phenomena were the rising or sinking of some brilliant stars. These phenomena appear every year at a fixed date in the sky. The annual occurrence of these events during a solar year has contributed to use them in order to organize practical-social activities, such as agriculture and navigation [15]. Hesiod, in the "Works and Days", mentions the harvesting period as the season when the constellation of Pleiades appears for the first time in the sky and as time of plowing, the time after the temporary disappearance of the Pleiades. The grape harvest must take place when Arcturus appears in the sky for the first time. Such phenomena and the date on which they occur are predicted using the Parapegma on the Mechanism [15].

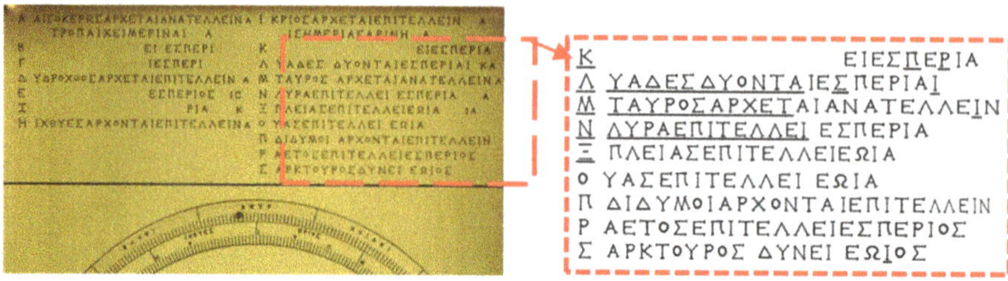

Figure 13. Part of the Parapegma of the Mechanism.

The rotation of any of the gears or any of the pointers results in simultaneous movement in all gears, and thus in the seven pointers, which show various astronomical phenomena in related mathematical scales. Based on an analysis of the applied torque of the operator while handling the Mechanism, and the necessity of precisely positioning the pointers, e.g., the Sun/Date pointer on the front of the Mechanism, the most probable scenario for the handling of the Antikythera Mechanism is the rotation of the Moon pointer [19]. By rotating the moon pointer and choosing a date, the operator can observe the astronomical phenomena occurs on that day. By choosing an astronomical phenomenon, he can observe the date on which it occurs.

2.4. Place and Date of the Construction of the Antikythera Mechanism

The Mechanism was possibly built in Rhodes, during the second century BC. Earlier evidence of similar machines report on Archimedes (287–212 BC) as a constructor of devices, which were depicted the celestial bodies [1,2]. The Mechanism cannot be built later than the coin-based shipwreck date within a few years (86 and 67 BC).The inscription letter style suggested its construction in the period 50–200 BC. An important limit could be set if we understood when the parameters required for the pin- and-slot lunar anomaly Mechanism were first deduced. It is known through Ptolemy that Hipparchus (190–120 BC) did characterize and quantify the anomaly by epicyclic and eccentric models of the lunar orbit. The manufacture took place after 170 BC and required that Hipparchus's values were involved [1].

The optimum latitude for fitting the astronomical phenomena listed in the parapegma on the Mechanism is consistent with the mid-Mediterranean, being around 35 degrees [15]. Rhodes (36 degrees) remains the most likely candidate. The Antikythera ship may have called there before the wreck, as it was known as a highly technological naval port with a thriving bronze industry, it was home to Hipparchus, it is the place for which we have a record of sighting of Mechanism with comparable functions and it might explain the presence on the Games' dial on the Mechanism of the Halieia Games, held in Rhodes [2].Another connection of Rhodes with the construction of similar Mechanisms is Cicero's reference, that he visited the laboratory of the astronomer Poseidonius (135–51 BC) in Rhodes, where he admired a celestial sphere made by Poseidonius [2].

The Metonic Cycle (upper back dial) contained a full calendar (repeated 19 times). Comparing this calendar with the calendars of the ancient Greek cities, it was found that it coincided with the calendars of the cities of Kerkyra, Vouthrotos and Dodona (in NW Greece) and Tavromenion (in Sicily) [15,16]. No significant coincidence with the survived calendars of other Greek cities was found. This could indicate that the Antikythera Mechanism was used, but not, necessarily, constructed in NW Greece?

2.5. Construction's Materials of the Mechanism

The material used to make the various parts of the Mechanism, except for its wooden mounting box, is bronze, a copper tin alloy. 1970–1974 chemical analysis was performed to determine the chemical composition of the metal alloy used in the manufacture of the Mechanism. The chemical analysis was performed by the spectroscopic method. Two small fragments of the Mechanism were studied. The chemical analysis showed that they were made of bronze, with a tin content of about 5% [3]. Newer analyzes by Panagiotis Mitropoulos in 2018 revealed three alloys, the main components of which are copper, tin and lead [3]. The shares of copper tin and lead varied. In any case, it can be assumed that the individual parts consist of different copper alloys from case to case [32].

2.6. Documentation of the Existents of the Necessary Knowledge for the Mechanism Design, Construction and Manufacturing at the Time and at the Place of Its Construction

2.6.1. Scientific and Technological Knowledge Necessary for the Design of the Mechanism

A question, which arises, is whether the Greek astronomers 2200 years ago, had the necessary knowledge for the calculation of the astronomical phenomena, calculated by the Mechanism? Various sources show that they had them [21].

In ancient Greece there was a solar calendar and a lunar calendar. The difference between then was 11 days. Meton (fifth century BC) connected the moon calendar to the annual. He calculated that about 19 solar years include about 235 synodic lunar months [2]. Callipus (370–300 BC) has calculated that Meton was making a mistake one day every 76 years [2]. These calculations are performed by the Mechanism and appear on two scales at the back side of the Mechanism. The Callippic pointer of the subsidiary dial within the Meton scale (upper back spiral of the Mechanism) indicated when the correction must take place. Aristarchus (310~230 BC) was the first astronomer in the history, who talk about the heliocentric system and not Copernicus [17]. He calculated among others the size of Earth, Sun and Moon. Hipparchus [17] (190–120 BC) is considered as the greatest astronomer of all time. He lived in the time of the construction of the Mechanism in Rhodes, where probably the Mechanism was built. He calculated, among others, the transient motion of the Earth (precession of the equinoxes) which lasts 25,800 years (see Section 4.1). Another calculation of him is the determination of the distance between Earth, the Sun and Earth and Moon.

2.6.2. Technological Knowledge Necessary for the Manufacturing of the Mechanism

Another question, which arises, is whether the ancient Greek mechanicians had the necessary technological knowledge for the manufacturing of the Mechanism. Various sources show that they had them. Until the discovery of the Antikythera Mechanism the construction of the first real gears was dated centuries later. Aristotle describes the rotation of cylindrical objects by friction, due to the roughness of their cylindrical surfaces. The creation of surfaces with higher roughness slowly led to the development of teeth and gears. The gears of the Antikythera Mechanism (second century BC) are the first in world history. Two references related to calculations and constructions of gears are from Heron of Alexandria (10–70 AD) and Pappus of Alexandria (290–350 AD)

Heron of Alexandria was an engineer and geometer. He lived in Alexandria, Egypt. In his description of the construction of an odometer he mentions and describes gears. Pappus of Alexandria describes several machines that were described by earlier mathematicians and engineers, such as Archimedes, Heron, etc. and included gears [33]. For two gears to work together, they must have the same ratios of diameter to number of teeth. This relationship is called today "module". From Papus's writings, it is clear, without any doubt that the Greeks knew the module in antiquity. Very likely, the gears of the Mechanism were made of cold forged thin bronze plates by sawing, removing redundant material and leveling with a hammer [32]. The history of gear technology has been postponed for many centuries.

The Mechanism includes, besides gears, complex geometry axles and shafts, as well as other metal parts. For their manufacturing, machine tools were required. Did the Greeks have machine tools and corresponding tools at the time of construction of the Mechanism?

The text of the inscription from the fourth century BC shown in Figure 14 concerns the construction of bronze axes" *Πόλος*" for the Filonian gallery in Eleusis, Greece using lathe. On this marble inscription is written among others " … *a copper alloy from Marion*

(Cyprus) must be used, consisting of 11 parts copper and one-part tin ... " This alloy is called bronze today. The parts of Antikythera Mechanism are made of bronze (see Section 2.5). Subsequently, it is written as *" ... Turn the axes according to the example ... "* This inscription shows that many years before the creation of the Mechanism, the Greeks had and used lathes. This is also apparent from other sources [17,32]. For the machining of bronze pieces, steel cutting tools are necessary. It follows from several sources that the Greeks at that time used such machine tools and cutting tools [17,32].

Figure 14. Marble inscription, fourth century BC. Archaeological Museum of Eleusis, Greece.

3. Construction of Digital 3D Representations and Reconstruction of Physical Functional Copies

From 1998, many animations and simulations of the Mechanism were created by M. Roumeliotis [34], which can be searched online. In relation to the physical models, from 1900 until today, many scientists have done research around reconstruction of the Mechanism. 1928 admiral I. Theofanidis with the contribution of E. Zinner, R.T. Gunther and W. Hartner, they listed the visible gears and circles from the back side of the Mechanism

and they characterized the Mechanism as an astrolabe. Admiral Theofanidis read 350 letters of the inscriptions and was the first who attempted to make a replica model of the Mechanism [3,17].

In 1985, A. Bromley continued the research of Price, including with respect to his remarks. On his first attempt, he worked with F. Percival to construct a replica of the Mechanism. Five years later in 1990, he worked with M. Wright, and they used X-ray Linear Motion Tomography, in order to define the array of gears in space. This method did not achieve the desired results. M. Wright claimed that the Mechanism could predict the movements of the Sun, the Moon and of five planets [27]. In 2005, he constructed a model using approximately 40 gears [7].

In 2008, a research team was created in the Aristotle University of Thessaloniki, which set as its main goal the construction of accurate and functional models of the ancient Mechanism. The first replica [11] was constructed using the mean data described in the publication «Decoding the ancient Greek astronomical calculator, known as the Antikythera Mechanism» [10]. This model was not functional, as the diameters of all gears were designed based on the mean values of the previously-mentioned article [10], with a tolerance of approximately 1 mm. This tolerance, in combination with the measured axial distances, led to non-functional gear trains, as some cooperating gears were blocked (due to larger diameters to the given axial distance) and others were not in contact (due to smaller diameters to the given axial distance).

To build a functional model, the parameters of the components of the Mechanism needed to be calculated with a higher accuracy than the mean values of the data of the aforementioned publication. Therefore, in 2009, two different algorithms were developed using FORTRAN code to make all these calculations [11,12].

The first algorithm uses the number of the teeth of each gear, as well as the root angle as constant variables, and calculates all the possible combinations of internal and external diameters. This is developed to find a ratio of the mean diameter and the number of teeth, which is common to all gears (similar to the "module" of the modern gears). Using the values of all gears from the first algorithm, the second calculates all the axial distances for the gears, taking into account the axial distances at the tomographies, as it had been measured from two different PhD fellows. Afterward, a new design was made, which included all the new data, and in 2011, five similar accurate and functional models were constructed (Figure 15) for museums and organizations, such as the National Archaeological Museum of Athens and the Musée des Arts et Métiers in Paris. From then until the present, many updated accurate and functional replicas in scale 1:1, as well as in scale 3:1 (Figure 16) have been constructed and exhibited in various museums and other institutions worldwide [17,18,35]. The updates refer to new letters and symbols of the inscriptions and some supporting mechanical components observed in the tomographies. The 1:1 model is made using the same materials as the original mechanism and represents the most accurate copy of the ancient device. The 3:1 model is made of bronze and Plexiglas, and is an educational model as the transparent materials reveal the internal of the device.

Figure 15. Accurate and functional model of the Mechanism scale 1:1 (research team of AUTH-2011).

Figure 16. Transparent educational and fully functional model of the Antikythera Mechanism in scale 3:1 (Research team of AUTH-2017).

4. Prediction Accuracy

4.1. Development of an Application to Calculate the Prediction's Accuracy of the Antikythera Mechanism

In order to verify the accuracy of the prediction of the Mechanism, an application has been developed in a virtual reality using Windows operating system. The indicators on both sides of the Mechanism are presented in an image format, rather than a numerical result. This application is also used to verify the accuracy of the predictions of the physical models developed by the research team of the Aristotle University of Thessaloniki. The application has been developed on Unity platform in 2D format. The dates and zodiac scales of the front side of the Mechanism are redesigned to correspond to the present. A peculiarity of the application, as well as the corresponding physical model is related to the scale of the zodiac signs.

Due to the passage of about 2150 years from the time of construction of the Mechanism, and due to the phenomenon of the precession of the equinoxes, the zodiac constellations have rotated by about 30 degrees.

The precession of the equinoxes is the motion of the equinoxes along the ecliptic, caused by the cyclic precession of Earth's axis of rotation with a period of 25,772 years. This means that 25,772 years correspond to 360 degree of precession of the equinoxes. If we consider that the Mechanism was built around 150 BC, it means that 2170 years have passed. Therefore, the equinoxes, and thus the Zodiac has shifted by 30 degrees $(2170 \times 360)/25{,}772 = 30.1$ degrees). For this reason, the scale of the signs was redesigned to reflect the current reality. The phenomenon of the earth's transient movement (see Figure 17) was known in antiquity, and Hipparchus was the first to calculate the movement. He is considered the father of astronomy, and most likely the maker of the Mechanism (see Section 2.4).

The implementation of the application's functions has been developed in C# programing language, which is compatible with Unity platform. By selecting a date, the data are exported to the date pointer, which shows the selected date on the Mechanism. The rest of the indicators demonstrate the astronomical phenomena that will take place that day. To verify the accuracy of the results, it is possible to compare the results for each prediction with corresponding predictions on NASA's website via a button, which includes the NASA website link. The application consists of two screens. The first screen contains the basic functions of an application, such as the "Play", "Help", "About Us" and "Quit" buttons. The Help button contains the necessary assistance for the correct interaction with the application. The functionality of the application become more understandable. The "Quit" category gives the command to exit the application. When the Play button is activated, the main screen opens, which contains the two sides of the Mechanism with the corresponding indicators and the calendar where the date of prediction of the phenomena is selected. By selecting the date and pressing the "Calculate" button, the Mechanism predicts the phenomena that play a role on the selected date. To decode the symbols of the Saros spiral dial (eclipses scale), further help is provided via an option.

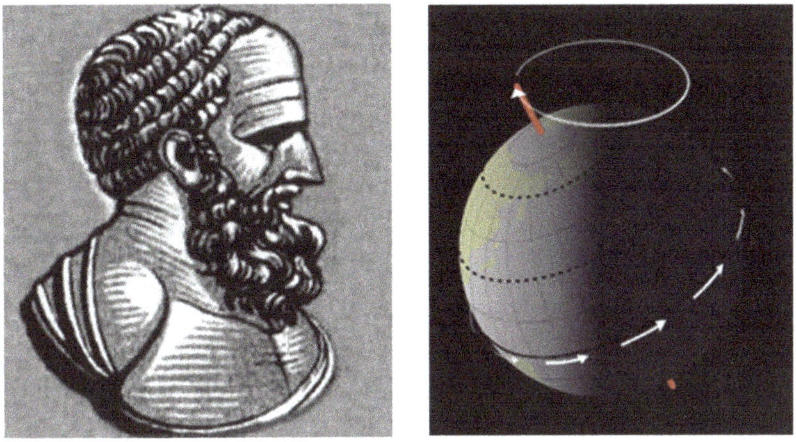

Figure 17. Hipparchus (**left**), the transient motion of the Earth, which lasts 25772 years (**right**).

In order to verify the correction of the application, important information, such as the calculation of the leap years that the Mechanism took into account were also considered when implementing the application. Apparently, in leap years that occur every four years, the extra day does not appear in the calendar of the Mechanism. The subdivisions of the date dial on the front side of the Mechanism are 365, and as a result it is not possible to indicate the extra day every four years (29 February). This feature is integrated into the application, which calculates the leap years that have elapsed from the initial settings of the Mechanism and gives the value of the variable that counts these years, as a rotation of the date ring by as many holes as the leap years, thus considering these years for the correct indication of the Sun/Day pointer (see Section 2.1.1 and Figure 6).

To verify the accuracy of the results, it is possible to compare the results for each prediction with corresponding predictions on NASA's website (https://eclipse.gsfc.nasa.gov/, accessed on 5 August 2021) via a button with the NASA website link.

4.2. Examples of the Application's Predictions

4.2.1. 1st Example: Prediction of the Astronomical Phenomena for the 26 May 2021

Figure 18a shows the results of the application for the 26 May 2021 which are: Sun is projected on the constellation of Taurus, Moon is projected on the constellation of Libra, Full Moon and Moon Eclipse. In relation to the Lunar Eclipse, the results are explained in Figure 18b. Figure 18c illustrates the corresponding data from the NASA webpage for the same day (26 May 2021), in order to verify the forecasts of the Mechanism.

(**a**)

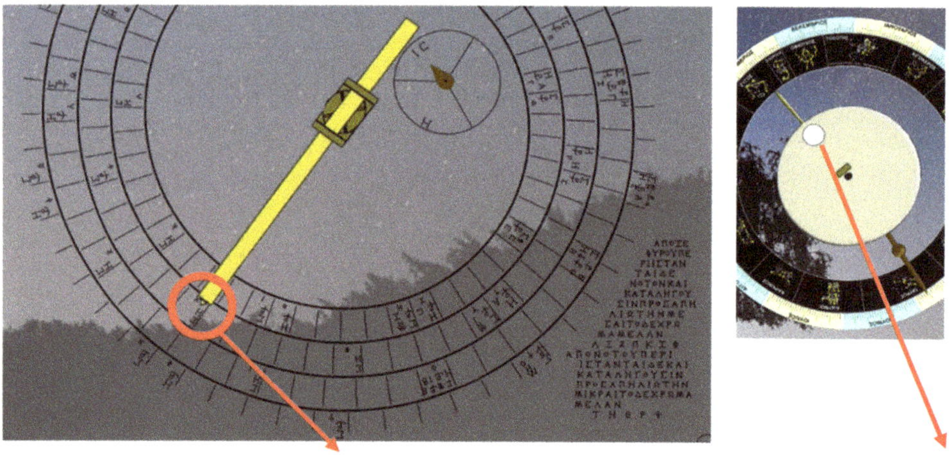

Moon Eclipse (Σ = Σελήνη = Moon) or Sun Eclipse (H = Ἥλιος = Sun) → **Moon Eclipse** ← Full Moon

(**b**)

Lunar Eclipses: 2021 - 2030

Calendar Date	TD of Greatest Eclipse	Eclipse Type	Saros Series	Umbral Magnitude	Eclipse Duration	Geographic Region of Eclipse Visibility
2021 May 26	11:19:53	Total	121	1.009	03h07m **00h15m**	e Asia, Australia, Pacific, Americas

(**c**)

Figure 18. (**a**) Astronomical phenomena on 26 May 2021. (**b**) The Moon Eclipse on 26 May 2021. (**c**) NASA Website: Astronomical phenomena dated 26 May 2021 (Full Moon, Moon Eclipse).

4.2.2. Second Example: Prediction of the Astronomical Phenomena for 10 June 2021

Figure 19a shows the results of the application for the 10 June 2021, which are: Sun and Moon projected on the constellation of Taurus, New Moon and Sun Eclipse.

In relation to the Solar Eclipse, the results are explain in Figure 19b. In Figure 19c, the corresponding data from the NASA webpage for the same day (10 June 2021).

(**a**)

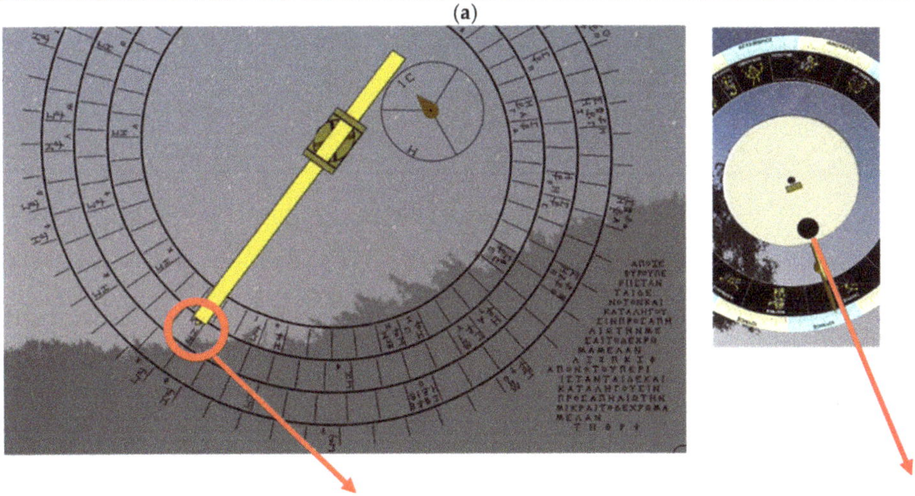

Moon Eclipse (Σ = Σελήνη = Moon) or Sun Eclipse (H = Ήλιος = Sun) ⟶ Sun Eclipse ⟵ New Moon

(**b**)

Figure 19. *Cont.*

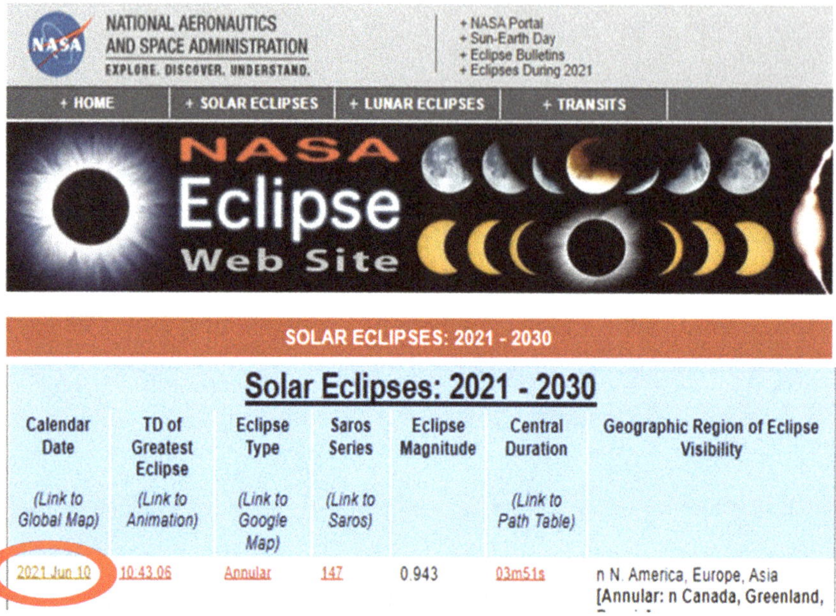

(c)

Figure 19. (**a**). Astronomical phenomena on the 10 June 2021. (**b**) The Sun Eclipse on 10 June 2021. (**c**) NASA Website: Astronomical phenomena on 10 June 2021 (New Moon and Solar Eclipse).

More predictions of the Mechanism and those corresponding from the NASA webpage are listed in Table 2.

Table 2. Various predictions of the Mechanism and the corresponding results from the NASA webpage.

Date	Position (Constellation)		Astronomical Phenomena				NASA Predictions(NASA SKYCAL Sky Events Calendar)
	Sun	Moon	New Moon	Full Moon	Sun Eclipse	Moon Eclipse	
12 October 1977	Virgo	Virgo	✓	✗	✓	✗	Solar Eclipse on 12 October 1977
19 August 2001	Leo	Leo	✓	✗	✗	✗	New Moon on 19 August 2001
11 October 2007	Virgo	Virgo	✓	✗	✗	✗	New Moon on 11 October 2007
29 December 2035	Sagittarius	Sagittarius	✓	✗	✗	✗	New Moon on 29 December 2035
17 January 2056	Sagittarius	Sagittarius	✓	✗	✓	✗	Solar Eclipse on 31 January 1999
8 February 1982	Capricorn	Leo	✗	✓	✗	✗	Full Moon on 8 February 1982
31 January 1999	Capricorn	Cancer	✗	✓	✗	✓	Moon Eclipse on 31 January 1999
26 May 2021	Taurus	Libra	✗	✓	✗	✓	Moon Eclipse on 26 May 2021
5 May 2023	Aries	Libra	✗	✓	✗	✓	Moon Eclipse on 5 May 2023
11 August 2071	Cancer	Capricorn	✗	✓	✗	✗	Full Moon on 11 August 2071

4.3. Calculate the Exact Date of the Celebration of Easter

According to the rules of the Christian church, Easter must be celebrated on the first Sunday after the first full Moon following the Spring Equinox. Since the date of the Easter celebration is related to the astronomical phenomenon of the lunar phases, which is calculated by the Mechanism, it is possible to calculate the exact date of the celebration of Easter without any intervention in the structure of the Mechanism. Based on this, the possibility of predicting the exact date of the celebration of Orthodox, Catholic and Jewish Easter has been developed in the application.

4.3.1. Predictions of the Date of the Catholic Easter of 2021

Figure 20a shows the first full Moon following the Spring Equinox of 2021 (Monday, 29 March 2021) based on Gregorian calendar, as it is calculated by the Antikythera Mechanism application. Figure 20b shows the date (4 April 2021) of Catholic Easter using the application.

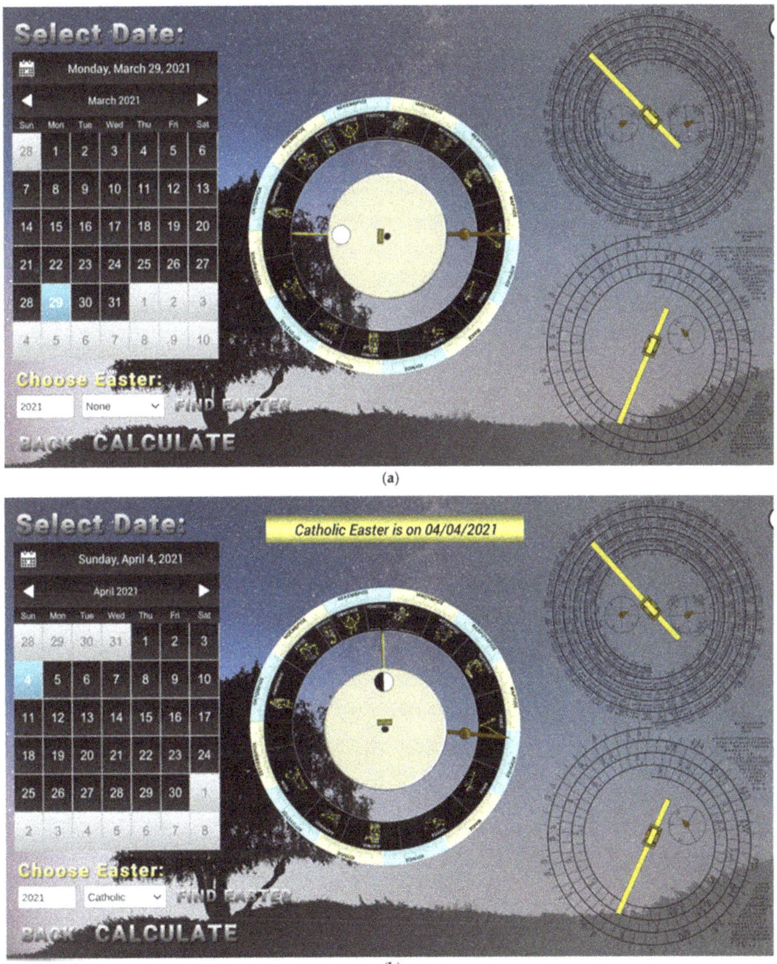

Figure 20. (**a**) The 1st full Moon following the Spring Equinox of 2021 based on Gregorian calendar. (**b**) Catholic Easter of 2021 (4 April 2021).

4.3.2. Predictions of the Orthodox Easter Date in 2021

Orthodox Easter follows the same rule, regarding the first full Moon following the Spring Equinox, but it is calculated based on the Julian calendar (old Calendar). Figure 21a shows the first full Moon following the Spring Equinox of 2021 (Tuesday, 27 April 2021) based on the Julian calendar, as it is calculated by the Antikythera Mechanism application and in Figure 21b the date (2 May 2021) of the Orthodox Easter following the application.

(a)

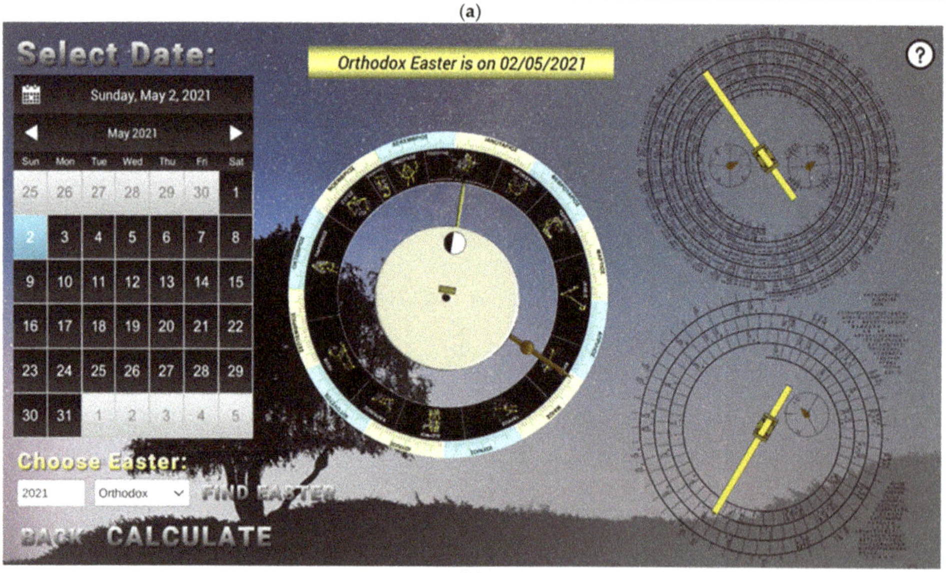

(b)

Figure 21. (**a**) The first full Moon following the Spring Equinox of 2021 based on the Julian calendar. (**b**) Orthodox Easter of 2021 (2 May 2021).

Table 3 provides some additional examples of the Easter date forecasts for different years from 2004 to 2044. For all the dates included in Table 3, the application calculates the same predictions.

Table 3. Catholic and Orthodox Easter predictions.

Year	Antikythera Mechanism Application		https://en.wikipedia.org/wiki/List_of_dates_for_Easter	
	Date of Celebration the Catholic Easter	Date of Celebration the Orthodox Easter	Date of Celebration the Catholic Easter	Celebration the Orthodox Easter
2004	April 11		April 11	
2009	April 12	April 19	April 12	April 19
2015	April 5	April 12	April 5	April 12
2020	April 12	April 19	April 12	April 19
2021	April 4	May 2	April 4	May 2
2022	April 17	April 24	April 17	April 24
2025	April 20		April 20	
2032	March 28	May 2	March 28	May 2
2044	April 17	April 24	April 17	April 24

4.4. Accuracy of Predictions of the Replica of the Antikythera Mechanism

Figure 22 shows a physical model, a replica of the Antikythera Mechanism, in which we have recorded on the front the current positions on the scales of dates and zodiac signs.

Figure 22. Displays on the front site of the Mechanism for 15 June 2011.

On 15 June 2011, the following astronomical phenomena occurred: The Sun was projected on the constellation of Taurus, the Moon was projected on the constellation of Sagittarius, there was a Full Moon and simultaneously there was a Lunar Eclipse at 23:00.

The corresponding indications of the replica are:

Front site (Figure 22):

15 June 2011, Sun in the constellation of Taurus, Moon in the constellation of Sagittarius and Full Moon. Thus, it is possible to observe a moon eclipse on this day.

Back site (Figure 23):

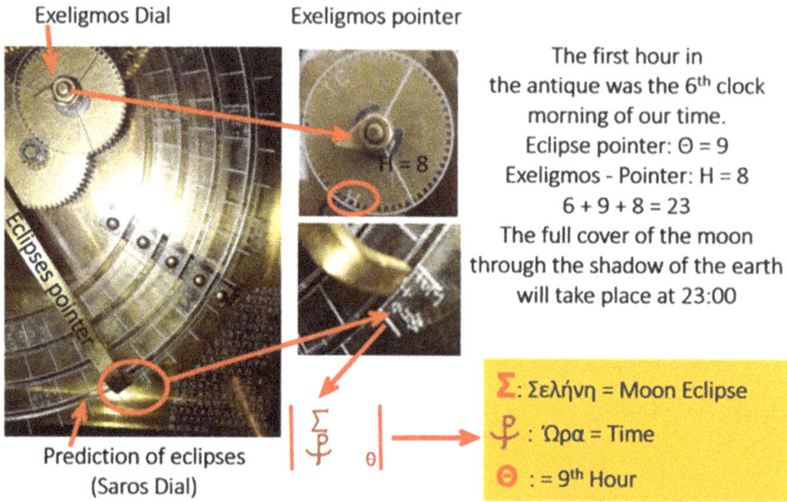

Figure 23. Displays on the back site of the Mechanism for 15 June 2011.

Observing the eclipse pointer on the back site of the Mechanism, a lunar eclipse will indeed occur. Specifically, regarding the lunar eclipse, the Mechanism showed:

(a) Eclipses pointer: Σ (Σελήνη = Moon), Θ = 9 (time of the event 9 o'clock); and
(b) Exeligmos Index: H = 8 h delay (must be added to Θ).

Considering that the first hour in antiquity was the 6th hour of the present era we have Θ + H + 6 = 9 + 8 + 6 = 23. That means the full coverage of the moon by the shadow of the Earth will take place at 23:00.

As it turns out, the predictions of the Antikythera Mechanism exactly coincide with the phenomena that actually occurred on 15 June 2011.

4.5. Example of Calculations That Simulate the Kinematics of the Mechanism

The transmission rate (GTR) between two gears is calculated by:

$$\text{GTR1, 2 (between the drive gear 1 and the driven gear 2)} = \frac{\text{Z1 (Number of teeth of gear 1)}}{\text{Z2 (Number of teeth of gear 2)}}$$

In Figure 24, we observe the mechanical flow diagram of the gear train that starts from gear b_2 (mounted on the b_{out} axis), which moves the Sun/Date pointer and leads to the gear b_3 (mounted on the b_{in} axis), which rotates the Moon pointer.

To identify the Moon pointer's angle of rotation, for a given rotation of the Sun pointer, we need to know the transmission rate of the gear train shown in the Figure.

Based on the flow diagram in Figure 24, the whole transmission rate begins from gear b_3, which rotates the shaft b_{in}, and the lunar pointer until the gear b_2, which rotates the shaft b_{out} and thus the Sun/Date pointer is:

$$GTR_{b(3),b(2)} = (Z_{b(3)}/\ Z_{e(1)})\ (Z_{e(6)}/\ Z_{k(2)})\ (Z_{k(1)}/\ Z_{e(5)})\ (Z_{e(2)}/\ Z_{d(2)})\ (Zd_{(1)}/\ Zc_{(2)})\ (Z_{c(1)}/\ Z_{b(2)})$$

$$GTR_{b(3),b(2)} = \frac{32}{32} \times \frac{50}{50} \times \frac{50}{50} \times \frac{32}{127} \times \frac{24}{48} \times \frac{38}{64} = 0.0748$$

That means that for one rotation of the lunar pointer, the Sun/Date pointer will make 0.0748 rotations, which corresponds to 26.928 degrees or to 27.3 days. This is how long it takes for the Moon to complete one full rotation around Earth.

Therefore, given any fixed rotation value of an indicator (or a gear), one could reliably predict how much each gear and indicator would rotate.

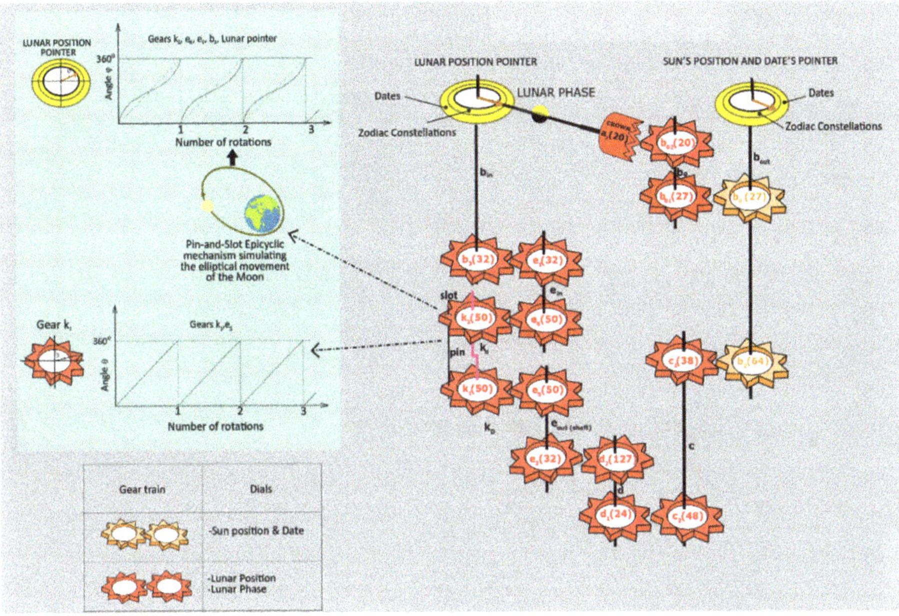

Figure 24. The gear train connecting the Sun/Date pointer and the Moon pointer.

In Figure 25, we observe the mechanical flow diagram of the gear train that begins from gear b_2 (mounted on the b_{out} axis), moves the Sun/Date pointer and leads to eclipses and exeligmos pointers.

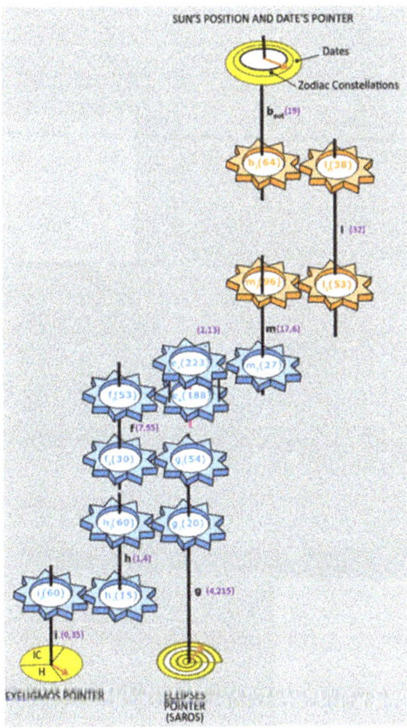

Figure 25. The gear train connecting the Sun/Date pointer and the eclipses and exeligmos pointers.

The application calculations are based on this logic. The corresponding part of the code, developed in the application for the calculations of the transmission rate of the gear train, are shown in Figure 25 and presented in Figure 26.

```
public void ExeligmosPointerDegree(double exeligmosDegree)
{
    double[] exeligmosPointerGears1 = {64, 53, 27, 188, 30, 20, 15};
    double[] exeligmosPointerGears2 = {38, 96, 223, 53, 54, 60, 60};
    double totalExeligmos = 0.0;
    double totalSumExeligmos = 1.0;
    double exeligmosAddedDegrees = 0.0;

    //Calculate gears ratio
    for(int i = 0; i < 7; i++)
    {
        totalExeligmos = exeligmosPointerGears1[i] / exeligmosPointerGears2[i];
        totalSumExeligmos *= totalExeligmos;
    }

    exeligmosAddedDegrees = (totalSumExeligmos * exeligmosDegree) + pointerExeligmos;
    pointerExeligmos = exeligmosAddedDegrees % 360;

    float vOut4 = ToSingle(pointerExeligmos);
    ExeligmosPointer.transform.rotation = Quaternion.Euler(0, 0, -vOut4);
}
```

Figure 26. Part of the code for the calculations of the transmission rate of the gear train, as shown in Figure 25.

4.6. *Accuracy of the Mechanism's Predictions*

The important elements that have to be considered for accurate forecasts of the Mechanism are:

(a) the accuracy of the equinoxes;
(b) the inaccuracy of the mechanism (mainly of the Saros scale) for in-depth predictions; and
(c) the experience of the operator in using the Mechanism and his knowledge of the astronomical phenomena.

4.6.1. Physical Accurate Functional Models (and Original Mechanism)

Handling Steps for Accurate Prediction of Future Phenomena

The handling steps for accurate prediction of future phenomena are as follows:

A. Rotate the gears until all the pointers move simultaneously. This will lead to the cooperating gears to be engaged, eliminating the gaps that occur due to the triangular shape of their teeth, or possible inaccuracies during their construction.
B. Adjustment of the indicator's position of the Mechanism, using data of a recent eclipse (within a period of about 100 years) and prediction of the future phenomenon.
C. If the next forecast refers to a date later than the previous one, no further action is required by the operator.
D. If the next forecast refers to a date earlier than the previous one, the Mechanism should be firstly turned backwards to a date earlier than the desired one and then forward to the desired date. By doing this, all the gaps that may occur, due to their opposite rotation, are eliminated.

Handling Steps for Accurate Prediction of Past Phenomena

The steps are the same, but with a different direction of rotation in each steps A, B, C and D.

4.6.2. Prediction with the Developed Application which Simulates the Mechanism

As in the computer application, the gears are considered ideal (without inaccuracies and gaps), all that is needed is to set the initial position of the indicators of the Mechanism, using the data of a recent eclipse (within a period of about 100 years). This can easily be done through this application.

5. Conclusions

One of the questions that concerns the international scientific research community is whether the Antikythera Mechanism was a scientific instrument for the safe prediction of astronomical phenomena or an educational tool for the schools of astronomy of antiquity. Based on the main goal of answering this question, but also to make the mechanism more accessible to the general public, we have developed an application, which simulates the exact operation of the physical model of the Mechanism. From all the tests that have been performed, it is concluded that the Mechanism accurately predicts the astronomical phenomena of the future and confirms with astonishing accuracy the astronomical phenomena of the past. The accuracy of the results of each prediction was determined by comparing its results with corresponding predictions on the NASA website. In combination with a physical functional model of the Mechanism, the application helped us to prove that the Antikythera Mechanism was a scientific instrument, the first computer in world history, which accurately predict astronomical phenomena.

It is important to emphasize that, from the study of the Mechanism, its usefulness in issues related to the present era have emerged. For example, it can be used to predict the dates of mobile church holidays, such as Easter. As the date of the celebration of Easter is related to the astronomical phenomenon of lunar phases, whereby the Mechanism calculates anyway, it is possible to calculate the exact date of the celebration of Easter without any intervention in its structure. Based on this, the possibility of predicting the exact date of the celebration of Orthodox, Catholic and Jewish Easter has been developed in the application. The results of all the tests proved to be accurate.

Author Contributions: Conceptualization, K.E. and M.E.; methodology, K.E., A.B. and M.E.; software, M.E. and N.K.; validation, A.B. and M.E.; formal analysis, M.E. and A.B.; investigation, K.E., M.E. and A.B.; resources, M.E., K.E. and N.K.; writing—original draft preparation, K.E.; writing—review and editing, K.E.; visualization, A.B. and M.E.; supervision, K.E.; project administration, K.E. All authors have read and agreed to the published version of the manuscript.

Funding: This research received no external funding.

Institutional Review Board Statement: Not applicable.

Informed Consent Statement: Not applicable.

Data Availability Statement: Data available on request due to restrictions. The data presented in this study are available on request from the corresponding author. The data are not publicly available due to privacy and ethical concerns and Intellectual Property Rights (application is yet to be released).

Acknowledgments: The authors acknowledge John Seiradakis (University of Thessaloniki), Mike Edmunds (Cardiff University) and Xenophon Moussas, holders of the study permit of the Mechanism from the Ministry of Culture of Greece, the Antikythera Mechanism Research Project for their contribution to the investigation of the Mechanism fragments, the AUTh Antikythera Mechanism Research Team for their support during this research and the National Archaeological Museum of Athens for their cooperation.

Conflicts of Interest: The authors declare no conflict of interest.

References

1. Seiradakis, J.H.; Edmunds, M. Our current knowledge of the Antikythera Mechanism. *Nat. Astron.* **2018**, *2*, 35–42. [CrossRef]
2. Jones, A. *A portable Cosmos*; Oxford University: New York, NY, USA, 2017.
3. Kaltsas, N.; Vlachoyanni, H.; Bouyia, P. *The Antikythera Shipwreck*; National Archaeological Museum of Athens: Athens, Greece, 2012. (In Greek)
4. Zafeiropoulou, M. The Antikythera Thesaurus. In Proceedings of the Oral presentation at Symi Festival, Symi, Greece, 31 August 2007.
5. *Archaeological Ephemeris*, Issue 1 &2, 3rd Period. 1902; 145–173.
6. Price de Solla, D. Gears from the Greeks: The Antikythera Mechanism—A calendar computer from ca 80 BC. *Trans. Am. Phil. Soc. New Ser.* **1974**, *64*, 1–70. [CrossRef]
7. Wright, M.T. Epicyclic Gearing and the Antikythera Mechanism, part 2. *Antiqu. Horol.* **2005**, *29*, 51–63.
8. Ramsey, A. The latest techniques reveal the earliest technology – A new inspection of the Antikythera Mechanism. In Proceedings of the International Symposium on Digital industrial Radiology and Computed Tomography, Lyon, France, 25–27 June 2007.
9. Malzbender, T.; Gelb, D.; Wolters, H. Polynomial Texture Maps. Available online: http://www.hpl.hp.com/research/ptm/papers/ptm.pdf (accessed on 15 August 2021).
10. Freeth, T.; Bitsakis, Y.; Moussas, X.; Seiradakis, J.H.; Tselikas, A.; Mangou, H.; Zafeiropoulou, M.; Hadland, R.; Bate, D.; Ramsey, A.; et al. Decoding the ancient Greek astronomical calculator known as the Antikythera Mechanism. *Nature* **2006**, *444*, 587–591. [CrossRef] [PubMed]
11. Efstathiou, K.; Basiakoulis, A.; Efstathiou, M.; Anastasiou, M.; Seiradakis, J.H. Determination of the gears geometrical parameters necessary for the construction of an operational model of the Antikythera Mechanism. *Mech. Mach. Theory* **2012**, *52*, 219–231. [CrossRef]
12. Efstathiou, M.; Basiakoulis, A.; Efstathiou, K.; Anastasiou, M.; Boutbaras, P.; Seiradakis, J.H.S. The Reconstruction of the Antikythera Mechanism. *Int. J. Herit. Digit. Era* **2013**, *2*, 307–334. [CrossRef]
13. Anastasiou, M.; Seiradakis, J.; Carman, C.C.; Efstathiou, K. The Antikythera Mechanism: The structure of the mounting of the back-plate's pointer and the construction of the spirals. *J. Hist. Astron* **2014**, *45*, 418–441. [CrossRef]
14. Efstathiou, K.; Efstathiou, M.; Gearbox, C. The oldest known computer is a mechanism designed to calculate the location of the sun, moon, and planets (Cover Story) Mechanical Engineering. *ASME Magazine*, September 2018; 31–35.
15. Anastasiou, M.; Seiradakis, J.H.; Evans, J.; Drougou, S.; Efstathiou, K. The astronomical events of the Parapegma of the Antikythera Mechanism. *J. Hist. Astron.* **2013**, *44*, 125–138. [CrossRef]
16. Anastasiou, M. The Antikythera Mechanism: Astronomy and Technology in Ancient Greece. Ph.D. Thesis, Aristotle University of Thessaloniki, Thessaloniki, Greece, 2014. (In Greek).
17. Efstathiou, M. The usage of innovative techniques of 3d design, 3d scanning and 3d printing in the investigation of ancient artifacts and other objects so as, among others, to construct their accurate replicas—Case Study of The Antikythera Mechanism. Ph.D. Thesis, School of Mechanical Engineering, Aristotle University of Thessaloniki, Thessaloniki, Greece, 2018. (In Greek).
18. Efstathiou, M.; Skordaris, G.; Basiakoulis, A.; Efstathiou, K. Construction of accurate and operational models of the Antikythera Mechanism using various manufacturing techniques such as conventional cutting, laser cutting and 3D printing technologies. In Proceedings of the 6th International Conference on Manufacturing Engineering ICMEN, Thessaloniki, Greece, 5–6 October 2017; pp. 293–308.
19. Basiakoulis, A.; Efstathiou, M.; Efstathiou, K.; Skordaris, G.; Seiradakis, J.H. The handling of the Antikythera Mechanism. In Proceedings of the 6th International Conference on Manufacturing Engineering ICMEN, Thessaloniki, Greece, 5–6 October 2017; pp. 281–292.
20. Ptolemy III Euergetes. Macedonian king of Egypt. Available online: www.britannica.com/biography/Ptolemy-III-Euergetes (accessed on 15 August 2021).
21. Egyptian calendar. Available online: https://en.wikipedia.org/wiki/Egyptian_calendar, (accessed on 15 August 2021).
22. Star Sirius. Available online: https://en.wikipedia.org/wiki/Star_Sirius, (accessed on 15 August 2021).
23. Nut (goddess). Available online: https://en.wikipedia.org/wiki/Nut_(goddess) (accessed on 15 August 2021).
24. Parker, R.A. *The Calendars of Ancient Egypt*; The University Of Chicago Press: Chicago, IL, USA, 1950. Available online: https://oi.uchicago.edu/sites/oi.uchicago.edu/files/uploads/shared/docs/saoc26.pdf (accessed on 20 October 2021).
25. Theodosiou, S.; Danezis, M. *The Odyssey of Calendars*; Diablos Publications: Athens, Greece, 1995. (In Greek)
26. Spalinger, A. Ancient Egyptian Calendars. In *Handbook of Archeoastronomy and Ethnoastronomy*; Springer Science and Business Media: New York, NY, USA, 2015; p. 1489.
27. Wright, M.T. Epicyclic gearing and the Antikythera Mechanism, Part I. *Antiquar. Horol.* **2003**, *27*, 270–279.
28. Gourtsoyannis, E. Hipparchus vs. Ptolemy and the Antikythera Mechanism: Pin-slot device models lunar motion. *J. Adv. Space Res.* **2010**, *46*, 540–544. [CrossRef]
29. Voulgaris, A.; Christophoros, M.; Andreas, V.; George, B. Renumbering of the Antikythera mechanism saros cells, resulting from the saros spiral mechanical apokatastasis. *Mediterr. Archaeol. Archaeom.* **2021**, *21*, 107–128. [CrossRef]
30. Gourtsoyannis, E. *Science and Culture Promise, Challege and Demand 285–289*; Epikentro Publications: Thessaloniki, Greece, 2012.
31. Freeth, T.; Jones, A.; Steele, J.M.; Bitsakis, Y. Calendars with Olympiad display and eclipse prediction on the Antikythera Mechanism. *Nature* **2008**, *454*, 614–617. [CrossRef] [PubMed]

32. Vlachogianni, E.; Lagogianni-Georgakarakos, M.; Andrea, B.; Nikos Kaltsas, N. *Der Versunkene Shatz—Das Schiffswrack von Antikythera*; Andrea Bignasca: Basel, Switzerland, 2015.
33. Spandagos, E. *ΗΜαθηματική Συναγωγή του Πάππου του Αλεξανδρέως, vol Δ'*; Αίθρα: Athens, Greece, 2006. (In Greek)
34. Antikythera Mechanism Simulation. Available online: http://www.etl.uom.gr/mr/index.php?mypage=antikythera_sim (accessed on 20 October 2021).
35. The Antikythera Mechanism. The most accurate and functional replicas of the Antikythera Mechanism ever made! Available online: https://www.antimech.com (accessed on 20 October 2021).

heritage

Article

A Short Note on the Knossos Statuette Inscription

Len Gleeson

Independent Researcher, Melbourne, VIC 3915, Australia; len_gleeson@hotmail.com

Abstract: For the inscription of the Egyptian statuette in the Heraklion Archaeological Museum, the dedicator's second title has long been open to question. New and detailed physical evidence, based on optical profilometry, is presented here. The results show errors/omissions in the previously accepted reading and open the way to a much more plausible translation.

Keywords: optical profilometry; surface characterization; chromatic white light

Citation: Gleeson, L. A Short Note on the Knossos Statuette Inscription. *Heritage* **2021**, *4*, 3186–3192. https://doi.org/10.3390/heritage4040178

Academic Editor: Nicola Masini

Received: 13 August 2021
Accepted: 25 September 2021
Published: 8 October 2021

1. Introduction

The statuette fragment of an Egyptian official named User, which can be dated to the first half of the 12th Dynasty, was found by Evans at Knossos in Crete in 1900 [1]. Now accession no. 95 in the Heraklion Archaeological Museum, it is of 'diorite', or more correctly, Anorthosite Gneiss [2], indicating that User was an official of very high rank [3]. The hieroglyphic inscription has long presented serious problems with correct sign identification [4], no doubt due to their very small size and varying depths of engraving. Moreover, the stone is slightly translucent, and has its own random colouration and a poorly reflective surface. Simple visual inspection and basic photography are often inadequate or even misleading in such cases [5].

Modern optical profilometry equipment was employed to resolve these uncertainties, and although the results generally agree with the previously published hieroglyphic sign layout [6,7], some important deviations will be noted.

2. Materials and Methods

The statuette's engraved surfaces, which are nominally flat, were scanned by an optical profilometry system from Fries Research and Technology GmbH, using a chromatic white light (CWL) technique. The equipment used was a Fries Research and Technology MicroSpy® Mobile, on site at the Heraklion Archaeological Museum, May 2015.

3. Results

3.1. Right Side Inscription

The area below the Standard sign, and to the top-left of the Canine-headed sign was scanned, as shown in Figure 1a,b. The inset at the bottom left shows the area scanned with respect to the traditionally accepted sign layout. Note that, for these figures and other views, the vertical (depth) scale is enlarged for more detail.

There is a distinct feature present (at the arrows' intersection), albeit not so deep as the surrounding signs; it is roughly rectangular, with a well-defined lower horizontal edge.

This feature is just discernible from an early photograph [8], and the sharp lower edge is depicted in the museum's exhibition catalogue [9]. The feature also has very good positioning, both horizontally and vertically, in between the two neighbouring hieroglyphic signs. It includes multiple, very fine striations of approximately 0.25 mm width—see Figure 1c for a cross-section along line MM' through the feature.

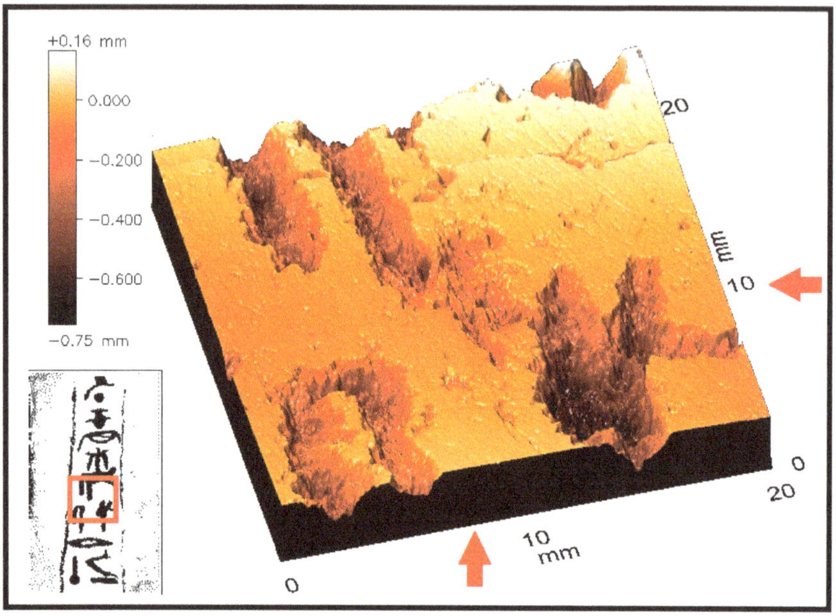

(**a**)

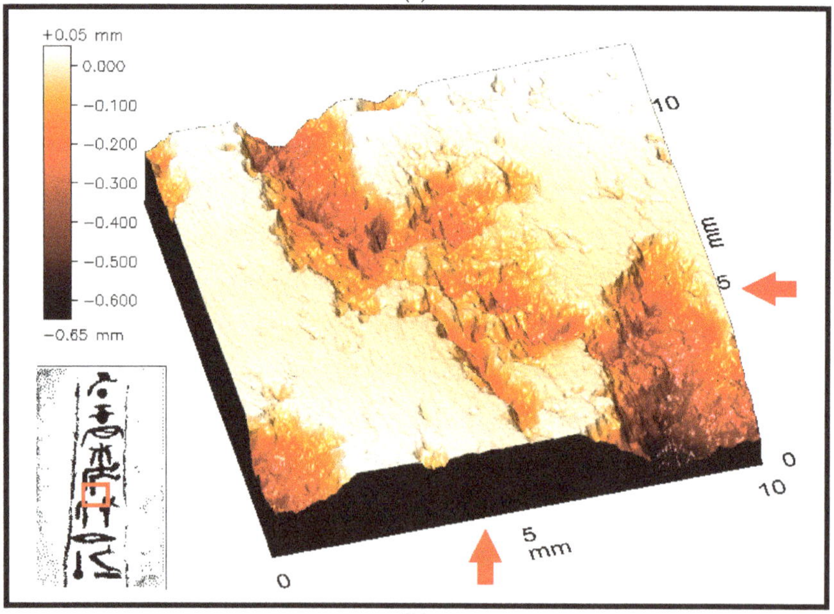

(**b**)

Figure 1. *Cont.*

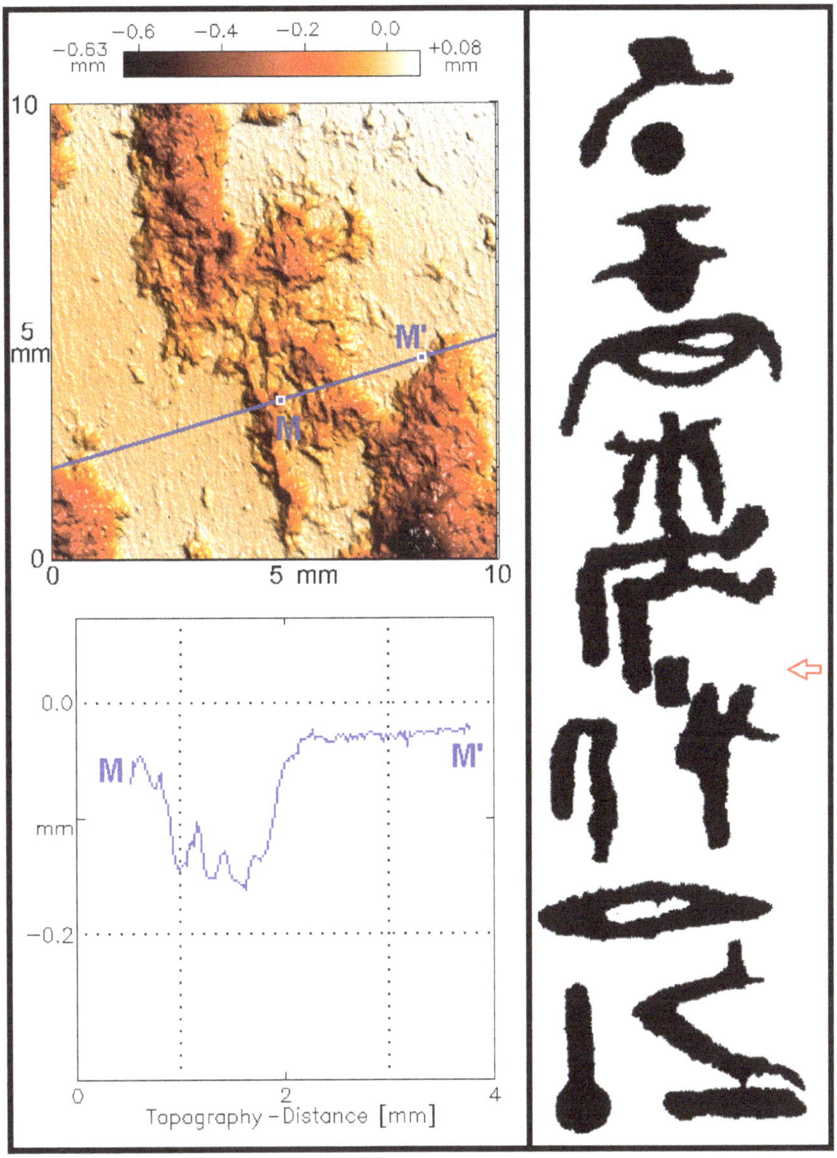

(**c**,**d**)

Figure 1. (**a**) Right side inscription, showing feature © Heraklion Archaeological Museum. (**b**) Right side inscription, higher magnification © Heraklion Archaeological Museum. (**c**) {above} Right side inscription, cross-section through feature. (**d**) {right} Revised Right side sign layout © Heraklion Archaeological Museum.

Such fine, parallel and uniformly spaced striations with a clear "V-shaped" cross-section, approximately aligned to the vertical axis, must surely be tool marks. All this strongly suggests a deliberately engraved sign and not something formed by later, random impact damage. This evidence would thus indicate the feature to be part of the inscription, clearly the Vertical Stroke, Gardiner's Z1. See Figure 1d for the revised sign layout of the statuette's right side.

3.2. Rear Inscription

User's second title, involving the Standard sign, is also found on the statuette's rear, so it is reasonable to expect a Vertical Stroke sign here as well. Although none is so far acknowledged, the traditional rear sign layout [6,7] does, however, suggests a less than ideal spacing for the Rear's left register, with a small but incongruous gap between the Standard sign and the first two signs of the name "User".

Optical profilometry on the statuette's rear, in Figure 2a, displays the area immediately below the Standard sign. Shown (at the intersection of the two arrows) is a small, shallow feature in line with the Canine-headed sign beneath. It consists of two roughly circular indentations, each approximately 0.1 mm deep and 0.6 mm diameter, and bridged by a faint, slightly curved line approximately 0.01 mm deep and 2.2 mm long. The centres of the circular indentations are aligned 4° clockwise with respect to the vertical axis of the inscription. See Figure 2b for a cross section along line NN′ through this feature.

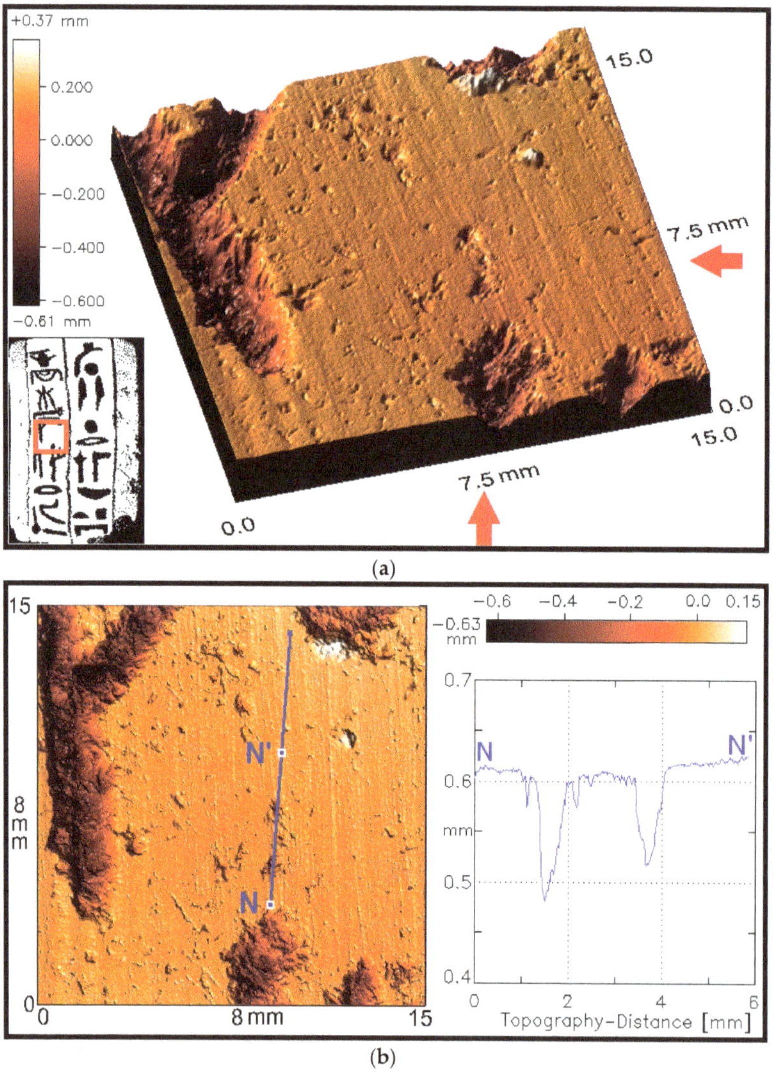

Figure 2. *Cont.*

(c)

Figure 2. (**a**) Rear inscription, showing feature © Heraklion Archaeological Museum. (**b**) Rear inscription, cross-section through feature © Heraklion Archaeological Museum. (**c**) Revised Rear sign layout © Heraklion Archaeological Museum.

It is proposed that this feature is part of an unfinished but *intended* engraving of a Vertical Stroke sign, on the basis of its two similarly shaped indentations, the shallow line between them—which does not overlap past either indentation—and its positioning and alignment compared to the acknowledged signs.

Like the right side inscription of Section 3.1, the Vertical Stroke sign was apparently largely overlooked and never properly engraved, presumably being finished with paint afterwards, a situation not unknown in Egyptology [10,11]. Refer to Figure 2c for the revised sign layout of the statuette's rear inscription.

4. Discussion—Proposed New Reading

4.1. Background

When the statuette was first unearthed in 1900, the signs immediately preceding the word 'User' were believed to be part of the dedicator's name [1]. It was only some years later that Griffith recognized that the dedicator was simply named User, a common Middle Kingdom name, and the preceding signs were "epithets, or a title" [12], a position later supported by Ward [13]. Even so, the Vessel signs were initially misread as Heart signs, so it was not until 1977 that Ward successfully translated User's first title as *Caster of Gold* [14].

4.2. Early Attempts

For User's second title, Griffith was of course thwarted by the humble optical facilities of the day and had no way of establishing the presence of the Vertical Stroke signs, as found in this study—Figures 1d and 2c. He proposed, in 1921, "whom the Wazet-nome (?) produced (?)" [12], at best a tentative first attempt, and openly doubtful. The concept was that the Wadjet nome 'produced' or 'begot' the person of User. However, on reflection this is quite implausible since one would expect virtually every official to be born in some nome or other, so that the title should be *very common* indeed, not rare or even unique. Note that Ward evidently rejected this translation, describing User's second title in 1961 as "completely obscure" [15].

4.3. *mś*

User's second title begins with Gardiner's F31 (fox tails), read as *mś*. As part of a title, it has two possible meanings: firstly, representing a "child" or "offspring", with the concept of "being given birth to" by a deity. The second possibility is a "worker", literally "one who shapes".

One can easily see how Griffith was misled and followed the former alternative, but now that User's first title, involving gold crafting, has finally been established, the evidence strongly favours the latter, as "worker" or "manufacturer". We have, for example "one who shapes hard stones", i.e., a stonemason [16], and "worker in precious stones" [17].

4.4. Helck

The interpretation of the *mś* sign as "worker" or "manufacturer" was apparently accepted by Helck, and he suggested, in 1979, that the second title might be "maker of tubular beads" [18]. However, a maker of tubular beads seems unlikely, since they were no doubt intended to be strung together to make jewellery items such as broad collars—a rather humble occupation, which cannot be seriously compared to a worker of molten gold, which was clearly a highly responsible, demanding and dangerous role. In any case, Helck's proposal offers no adequate explanation for the presence of the Standard sign together with the Snake sign.

4.5. Hayes

Hayes recognized [19] User's first title, involving gold casting, as well as the Wadjet Snake together with the Standard sign, and made the insightful and plausible suggestion *Maker of the Wadjet Standard*. That is, User was the artisan who constructed the Wadjet standard itself, probably an impressive, ornate object, quite likely of precious metals, perhaps used for some important religious or civil ceremonies. Note that other deity standards, as well as soldiers' standards, are indeed attested [20–22].

The new evidence established in this article, of Vertical Stroke determinative signs under the Standard signs, greatly supports Hayes' proposal by ensuring a logographic reading to the Snake/Standard sign combination. It should also be pointed out that these new signs were only discovered some years after Hayes' work.

5. Conclusions

State-of-the-art surface profiling equipment has been used on the Egyptian statuette, AM Heraklion no. 95, and has established previously unknown hieroglyphic signs, so that the dedicator's second title can now be read, with some confidence, as *Maker of the Wadjet Standard*.

Funding: This research received no external funding.

Acknowledgments: I acknowledge the vital assistance of all the staff of the Heraklion Archaeological Museum, Crete, particularly the Director, Stella Mandalaki, and the Curator Georgia Flouda. I am very grateful for the work of Barbara Richarz of Fries Research and Technology who obtained all measured data and images on-site at the Museum. I acknowledge the generous assistance of Grant Hayes of Macquarie University, Australia, with ancient Egyptian hieroglyphics. Thanks are also due to Simon Conner of the University of Liège, for assistance with the chronology of the statuette.

Conflicts of Interest: The author declares no conflict of interest.

References

1. Evans, A. The Palace of Knossos in its Egyptian Relations. In *Archaeological Report 1899–1900*; Griffith, F.L., Ed.; Egypt Exploration Fund: London, UK, 1900; pp. 65–66.
2. Aston, B.; Harrell, J.; Shaw, I. Stone. In *Ancient Egyptian Materials and Technology*; Nicholson, P., Shaw, I., Eds.; Cambridge University Press: Cambridge, UK, 2000; p. 33.
3. Wildung, D. Two representations of gods from the early Old Kingdom. *Misc. Wilbouriana* **1972**, *1*, 150f.
4. Evans, A. *The Palace of Minos at Knossos*; MacMillan & Co.: London, UK, 1921; Volume 1, p. 289, note 4.
5. Fischer, H.G. Varia Aegyptiaca. *JARCE* **1963**, *2*, 23. [CrossRef]
6. Edel, E. Die hieroglyphische Inschrift auf der dioritstatuette des User auf Knossos. In *Studies in Egyptology: Presented to Miriam Lichtheim*; Israelit-Groll, S., Ed.; Magnes Press: Jerusalem, Israel, 1990; p. 124.
7. Gill, D.; Padgham, J. 'One find of capital importance': A reassessment of the statue of User from Knossos. *Annu. BSA* **2005**, *100*, 45. [CrossRef]
8. Culican, W. *The First Merchant Venturers*; Thames & Hudson: London, UK, 1966.
9. Karetsou, A.; Andreadaki-Vlazaki, M. *Crete-Egypt Three Thousand Years of Cultural Links*; Hellenic Ministry of Culture: Heraklion, Greece, 2000; p. 62, upper-right drawing.
10. Bourriau, J. *Pharaohs and Mortals Egyptian Art in the Middle Kingdom*; Cambridge University Press: Cambridge, UK, 1988; pp. 30–31, for BM EA579.
11. Williams, C. *The Decoration of the Tomb of Per-Nēb*; Metropolitan Museum of Art: New York, NY, USA, 1932; plate II (b), (f) for MMA 26.9.1.
12. Evans, A. *The Palace of Minos at Knossos*; MacMillan & Co.: London, UK, 1921; Volume 1, p. 287.
13. Ward, W. *Egypt and the East Mediterranean World 2200–1900 BC*; American University of Beirut: Beirut, LB, USA, 1971; p. 81, note 333.
14. Ward, W. Lexicographical Miscellanies. *SAK* **1977**, *5*, 269–271.
15. Ward, W. Egypt and the East Mediterranean in the Early Second Millennium B. C. *Orientalia* **1961**, *30*, 28.
16. Erman, A.; Grapow, H. *Wörterbuch der ägyptische Sprache*; Akademie Verlag: Berlin, Germany, 1971; Volume 2, p. 138, no. 19.
17. Jones, D. *An Index of Ancient Egyptian Titles, Epithets and Phrases of the Old Kingdom*; Archaeopress: Oxford, UK, 2000; Volume 1, p. 451, no. 1689.
18. Helck, W. *Die Beziehungen Ägyptens und Vorderaisens zur Ägais bis ins 7. Jahrhundret v. Chr*; Wissenschaftliche Buchgesellschaft: Darmstadt, Germany, 1979; p. 47.
19. Hayes, G.; Macquarie University, Sydney, Australia. Personal communication, 2004.
20. Petrie, W. *Abydos*; Egypt Exploration Fund: London, UK, 1903; Volume 2, plate xxvii (top).
21. Faulkner, R. Egyptian Military Standards. *JEA* **1941**, *27*, 12–18.
22. Engelbach, R. Report on the Inspectorate of Upper Egypt from April 1920 to March 1921. *ASAE* **1921**, *21*, 73.

heritage

MDPI

Article

Identification of Dyes in Coptic Textiles from the Museum of Faculty of Archaeology, Cairo University

Ioannis Karapanagiotis [1,*], Omar Abdel-Kareem [2], Paraskevi Kamaterou [3] and Dimitrios Mantzouris [4]

[1] Department of Management and Conservation of Ecclesiastical Cultural Heritage Objects, University Ecclesiastical Academy of Thessaloniki, 54250 Thessaloniki, Greece
[2] Conservation Department, Faculty of Archaeology, Cairo University, Orman, Giza 12613, Egypt; omaa67@yahoo.com
[3] Department of Chemical Engineering, Aristotle University of Thessaloniki, 54124 Thessaloniki, Greece; Kamaterou@hotmail.com
[4] Independent Researcher, 17343 Athens, Greece; dimitrios.mantzouris@gmail.com
* Correspondence: y.karapanagiotis@aeath.gr

Abstract: High Performance Liquid Chromatography coupled to a Diode-Array-Detector (HPLC-DAD) is used to investigate samples which were extracted from ancient Egyptian textiles (4th–5th c. AD) of the Museum of Faculty of Archaeology, Cairo University. Madder is identified in several samples. According to semi-quantitative results, which are obtained from HPLC peak areas measured at 254 nm, madder that is rich in purpurin and poor in alizarin is identified in samples which were treated (i) only with madder and (ii) with madder and either indigo/woad (*Indigofera* species and other/*Isatis tinctoria* L.) or weld (*Reseda luteola* L.). The madder dye used in these samples could have been originated from *Rubia peregrina* L. However, the possible use of *Rubia tinctorum* L. (or other plants of the *Rubiaceae* family) by the Egyptian dyers cannot be ruled out, particularly if methods were developed by the ancient dyers to affect and control the relative composition of madder dye. The HPLC peak area ratio of alizarin versus purpurin is very high (>2.2) for samples which were treated with madder (probably originated from *R. tinctorum*) and a tannin source. Finally, in some samples, only indigoid dyes (indigo/woad) are identified.

Keywords: textile; Egypt; Coptic; dye; madder; indigo; woad; weld; HPLC

Citation: Karapanagiotis, I.; Abdel-Kareem, O.; Kamaterou, P.; Mantzouris, D. Identification of Dyes in Coptic Textiles from the Museum of Faculty of Archaeology, Cairo University. *Heritage* **2021**, *4*, 3147–3156. https://doi.org/10.3390/heritage4040176

Academic Editor: Diego Tamburini

Received: 22 August 2021
Accepted: 4 October 2021
Published: 7 October 2021

1. Introduction

The Coptic textiles of the Museum of Faculty of Archaeology, Cairo University, have been obtained either through excavations that were carried out by the University or as gifts given by the Arab Antiquities House (Museum of Islamic Art), the Committee for the Preservation of Arab Antiquities and several individuals. Examples of these valuable objects are shown in the photographs in Figure 1 [1]. The textiles of the Coptic collection correspond to different periods of the history of Egypt, and they were not subjected to dyestuff analysis in the past.

The origin of materials that were used by the Coptic dyers has been investigated in several previously published studies. Using modern analytical techniques, the following dyes have been identified in ancient Egyptian textiles: Armenian cochineal (*Porphyrophora hamelii* Brandt) [2–6], fustic (*Maclura tinctoria* L.) [7], henna (*Lawsonia inermis* L.) [7], indigoids (*Indigofera* species, *Isatis tinctoria* L. and other) [2–13], kermes (*Kermes vermilio* Planchon) [2,5,6], lac (*Kerria lacca* Kerr) [2–6,12,13], madder (*Rubiaceae* family) [2–13], mollusc purple (*Muricidae* family) [5,6,10,13–17], *Rhamnus* species [4], safflower (*Carthamus tinctorius* L.) [5], soluble redwood (*Caesalpinia* trees) [5,12], tannins [2,4,5,8,10] and weld (*Reseda luteola* L.) [2–13]. Moreover, a dye which should be similar to American annatto (*Bixa orellana* L.) has been included in the results of ancient Egyptian textiles, reported by Verhecken [5].

Figure 1. Examples of textiles from the Museum of Faculty of Archaeology, Cairo University.

Identification of dyes in textiles of the cultural heritage is commonly achieved by using High Pressure Liquid Chromatography coupled to a Diode-Array-Detector (HPLC-DAD) [18,19]. The method is employed herein to analyse samples which were extracted from twelve ancient Egyptian textiles of the Museum of Faculty of Archaeology, Cairo University. The HPLC-DAD results are analysed in order to discuss the possible biological origins of the madder dyes that were identified in the investigated samples. Therefore, the goal of the study is to add more data to the existing literature regarding the dyestuff sources used by the famous ancient Egyptian dyers.

Textile samples, investigated herein, are treated with methanolic solution of hydrochloric acid (HCl) to extract the dyes [20], which are finally dissolved in dimethyl sulfoxide (DMSO) and analysed by HPLC. In the harsh acidic conditions induced by the strong HCl, glycosides are hydrolysed to their parent aglycones. From a chemical point of view, this is clearly a disadvantage of the HCl method. However, when only very small samples from cultural heritage objects (e.g., paintings or even textiles) are available, then the aforementioned hydrolysis might be desirable to increase the chances to detect the important aglycone components. The latter often provide enough information to identify the biological source of a dye fixed on a historical or archaeological textile [21]. Several mild acids that retain glycosides have been suggested for dye extraction such as, for instance, formic [22–28], oxalic [23,25–27,29–32], hydrofluoric [25,33] and trifluoroacetic [23,27,34] acids. Other approaches for dye extraction suggested the use of ammonia-disodium EDTA to preserve glycosyl moieties [35] and an aqueous glucose solution for the extraction of madder [36]. However, regardless of the selected treatment method, the results obtained by

HPLC after the extraction of a dye from a textile do not coincide with the HPLC fingerprint of the biological (animal or plant) source. This is because the composition of a dye attached to a textile fibre is highly affected by the conditions of the dyeing process [37]. Moreover, the composition of a dyestuff source is affected by the geographical locations and growing environments [37].

2. Materials and Methods

2.1. Samples

Twelve (12) wool samples were extracted from Coptic textiles (4th–5th c. AD), which are preserved in the Museum of Faculty of Archaeology, in Cairo University. Photographs of some of the investigated textile samples are shown in Figure 2. In the Supplementary Materials, photographs of all the studied samples are provided. Samples were first studied using optical microscopy and were then analysed using High Pressure Liquid Chromatography coupled to a Diode-Array-Detector (HPLC-DAD) to identify the dyestuff materials. Identifications were achieved using HPLC-DAD data of standard materials, which are included in Table 1. Suppliers of the standard materials are described in Table 1. Indirubin and rubiadin were generously provided in pure forms by P. Magiatis (University of Athens) and R. Karadag (Istanbul Aydın University), respectively.

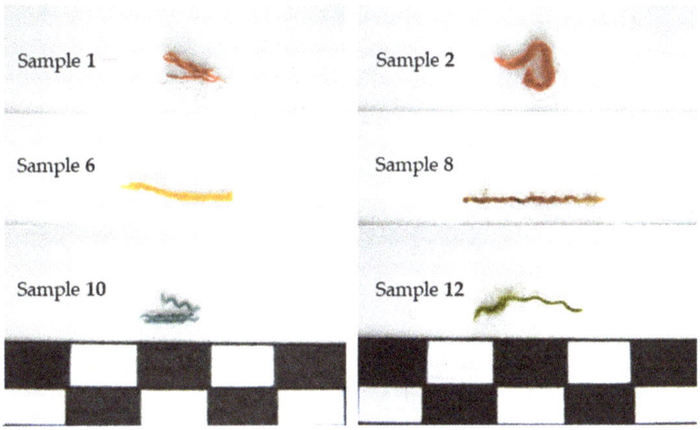

Figure 2. Examples of samples that were extracted from the Coptic textiles and included in the study. Rectangles of the ruler are 1 cm long.

Table 1. UV–Vis absorbance maxima of compounds which were identified in the investigated textile samples. Suppliers which provided the corresponding standard materials are described.

Compound	Supplier	Absorbance Maxima (nm)
Alizarin	Aldrich	248, 278, 429
Apigenin	Fluka	267, 336
Chrysoeriol	Extrasynthese	249, 267, 346
Ellagic acid	Fluka	253, 367
Indigotin	Fluka	243, 285, 606
Indirubin	P. Magiatis (gift)	243, 290, 364, 543
Luteolin	Sigma	253, 265, 345
Purpurin	Aldrich	255, 293, 479
Rubiadin	R. Karadag (gift)	244, 278, 412

2.2. Instrumentation

A Zeiss Axioskop 40 polarizing microscope was used to study the textile samples. Chromatography was carried out using an Ultimate 3000 (Dionex) system consisting of an LPG-3000 quaternary HPLC pump with a vacuum degasser, a WPS-3000SL autosampler, a column compartment TCC-3000SD and a UV–Vis Diode Array Detector (DAD-3000). The chromatographic separation was performed on an Alltima HP C18 (250 mm × 3 mm, i.d. 5 μm) Grace (Alltech Associates) column at a stable temperature of 35 °C. The injection volume was 20 μL.

2.3. Materials

The following chemicals were used to operate the chromatographic system: HPLC grade water (H_2O, ChemLab, Zedelgem, Belgium), HPLC grade acetonitrile (CH_3CN, ChemLab) and trifluoroacetic acid (TFA) (assay >99.0%, Sigma-Aldrich, St. Louis, MO, USA). Textile samples were treated using HPLC grade solvents of water (H_2O, ChemLab), methanol (MeOH, J.T. Baker, Radnor, PA, USA) and dimethyl sulfoxide (DMSO, Sigma-Aldrich) as well as pro-analysis grade hydrochloric acid (HCl) 37% m/m (Riedel-de Haën, Seelze, Germany).

2.4. Sample Treatment and HPLC Analysis

Textile samples were treated according to the HCl extraction method [20]. In particular, sample was immersed in a hot (100 °C) bath of 400 μL of H_2O:MeOH:37% HCl (1:1:2, *v*/*v*/*v*) for 10 min. The extract was then heated at 65 °C, and the liquid phase was evaporated under a gentle nitrogen flow. The precipitated dyes were dissolved in 200 μL of DMSO. This last step differs from the original method devised by Wouters [20], who suggested the use of H_2O and MeOH instead of DMSO used herein [38]. DMSO is superior compared to H_2O and MeOH for dissolving indigoid compounds [39–41]. The solution was centrifuged, and the upper liquid phase was immediately subjected to HPLC analysis. HPLC was operated by using a gradient elution program, which was previously developed for the analysis of natural dyes [42].

3. Results and Discussion

The UV–Vis absorbance maxima of the compounds which were identified in the extracts of the textile samples are summarized in Table 1. As shown in the table, compounds that are components of madder (alizarin, purpurin and rubiadin), weld (apigenin, luteolin and chrysoeriol) and indigoid dyes (indigotin and indirubin) were detected in the sample extracts. The HPLC results per sample are summarized in Table 2.

The discussion is organized as follows. First, the results achieved for the samples which were dyed with madder (i.e., samples 1–8) are reported and elaborated. Next, samples in which madder was not found (i.e., samples 9–12) are discussed.

According to the results of Table 2, madder was the most frequently identified dye, and it was found, in total, in eight (1–8) out of 12 samples included in the study. This result is not surprising, as madder has been often found in ancient Egyptian textiles [2–13] and has been used by the dyers since antiquity [21,43]. In four samples (1–4), madder was the only identified dye. These were red samples as shown, for instance, in the photographs of samples 1 and 2 in Figure 2. According to the results of Table 2, rubiadin was detected only in samples 1 and 4, whereas alizarin was found in the extracts of all four (1–4) samples. Apparently, purpurin, the marker compound for the identification of madder, was found in all four (1–4) samples.

Table 2. HPLC results for the studied textile samples: detected compounds and identified dyes.

Sample No	Colour	Detected Compounds	Identified Dyes
1	Red	Alizarin, Purpurin, Rubiadin	Madder
2	Red	Alizarin, Purpurin	Madder
3	Red	Alizarin, Purpurin	Madder
4	Red	Alizarin, Purpurin, Rubiadin	Madder
5	Greenish	Alizarin, Purpurin, Indigotin, Indirubin	Madder, Indigo/Woad
6	Yellow	Alizarin, Purpurin, Luteolin, Apigenin, Chrysoeriol	Madder, Weld
7	Red-brown	Alizarin, Purpurin, Ellagic acid	Madder, Tannin
8	Red-brown	Alizarin, Purpurin, Ellagic acid	Madder, Tannin
9	Blue-greenish	Indigotin, Indirubin	Indigo/Woad
10	Blue	Indigotin, Indirubin	Indigo/Woad
11	Blue	Indigotin, Indirubin	Indigo/ Woad
12	Green	Indigotin, Indirubin, Luteolin, Apigenin, Chrysoeriol	Indigo/Woad, Weld

In four other samples (5–8), madder was identified along with other dyes. In particular, in sample 5, madder was found in mixture with an indigoid dye, which can either be indigo or woad. Mixing madder and indigo/woad to imitate true, mollusc purple was a common practice in ancient Egypt [6]. However, purple threads in sample 5 were not visible by naked eye or under the microscope. Instead, the sample appeared to be rather greenish, as shown in Figure S1 of the Supplementary Materials. The chromatographic analysis of sample 6 is shown in Figure 3. The latter is provided as an example of the HPLC studies that were performed on the investigated samples. Apigenin, luteolin and chrysoeriol (weld components) were detected when the monitoring wavelength was set to 350 nm, whereas HPLC peaks corresponding to alizarin and purpurin (madder components) were revealed at a higher wavelength (450 nm). A direct comparison of the two HPLC profiles of Figure 3 should be carefully considered, as the two chromatograms were collected at different wavelengths. However, it can be observed that the peak areas of madder components are substantially lower than the corresponding areas of the weld components. The HPLC peak area ratio of luteolin versus purpurin, measured at 254 nm, was around two. Consequently, weld was the dominant dye in sample 6, and this is in agreement with the yellow colour of the sample, which is revealed in the photograph of Figure 2.

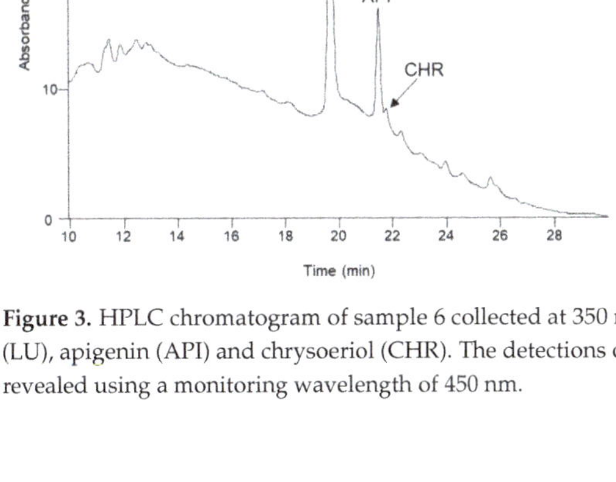

Figure 3. HPLC chromatogram of sample 6 collected at 350 nm showing the detections of luteolin (LU), apigenin (API) and chrysoeriol (CHR). The detections of alizarin (AL) and purpurin (PU) are revealed using a monitoring wavelength of 450 nm.

According to the results of Table 2, in the chromatograms of samples 7 and 8, madder components and ellagic acid (tannin) were detected. Tannins are very common in the flora; they possess interesting antimicrobial properties [44] and have been used both as dyes to produce black dyed fibres and as adhesives to promote the stabilization of colourants on textile substrates. The presence of tannin in samples 7 and 8 results in a dark hue compared to the usual red colour of madder. Hence, the colour of samples 7 and 8 appeared to be red-brown, as revealed, for instance, in the photograph of sample 8 (Figure 2) and in the Supplementary Materials (sample 7).

The term *madder* is used to describe different plants that belong to the *Rubiaceae* family. Important species which are often described in studies of objects of the Mediterranean cultural heritage are *Rubia tinctorum* L., *Rubia peregrina* L. and *Galium* species. *R. tinctorum* is found in Central and Southern Europe, Northern Africa and Central Asia, whereas *R. peregrina* (wild madder) has a range mostly restricted to Central and Southern Europe and Northern Africa [43,45,46]. *Galium* species can be found in any continent, whereas *Asperula tinctoria* L., which was probably also important in antiquity, is included in the native flora of Europe [43,45,46].

The HPLC results, collected from the analysis of samples extracted from cultural heritage objects, are often used to identify the exact plant species of madder dyes, lakes and paints [10,12,47–51]. In particular, the ratio of the HPLC peak areas of alizarin and purpurin was suggested as a criterion to identify *R. tinctorum* and *R. peregrina* [52]. It has been reported that *R. tinctorum* is rich in alizarin, whereas purpurin is the dominant anthraquinone in extracts of *R. peregrina* [52]. It should be noted, however, that purpurin is the main colouring compound in *Galium* and *Asperula* species [43]. Moreover, the large amount of rubiadin might be another marker for identifying *R. peregrina* [10,52], but more recent studies have raised concerns about the reliability of this criterion [49].

In the following, an attempt is carried out to discuss the possible biological sources of the madder dyes used in the studied textiles. The discussion focuses mainly on *R. tinctorum* and *R. peregrina*, which could have been used by the Egyptian dyers. It is stressed that the potential use of *Galium* and *Asperula* species cannot be ruled out. However, there are no chemical criteria (e.g., marker compounds or compositional data) that could be used to identify with confidence the use of these plant species in historical textiles. Figure 4 shows the ratios of the HPLC peak areas of alizarin versus purpurin (AL/PU) for samples 1–8. Peak areas were measured using HPLC chromatograms which were collected at 254 nm. This wavelength was used in the past to investigate *R. tinctorum* and *R. peregrina* samples [52], and it was, therefore, selected in the present study for comparative purposes. According to the results of Figure 4, madder dyes used for the treatment of the four samples (1–4) in which no other dye was found were poor in alizarin compared to purpurin. Likewise, small peak areas of alizarin compared to purpurin were measured for samples 5 and 6 in which madder was found in mixture with an indigoid dye and weld, respectively. It is reported that, for samples 1–6, the HPLC peak area ratio AL/PU measured at 254 nm ranged from 0.08 to 0.25. The low values of the AL/PU ratio indicate that alizarin was contained in samples 1–6 in small, but not negligible, amounts compared to purpurin. The results can offer support to the scenario that suggests that *R. peregrina* was used to treat samples 1–6; other possible sources, which are rich in purpurin, are of course the *Galium* and *Asperula* species. Moreover, as alizarin was detected in amounts that cannot be considered as negligible, the potential use of *R. tinctorum* cannot be excluded. Furthermore, it is possible that the Egyptian dyers had developed methods to affect the relative composition of madder plants before or during the dyeing process. For example, the relative composition of madder lakes which were prepared using *R. tinctorum* was tuned using materials and methods that were available in the past [53]. Control of the relative composition of the madder source provides control to the resulting colour of the final object. Madder rich in purpurin tends to induce a red-crimson hue [43] to the textile, which could be have been desirable by the ancient Egyptian dyers.

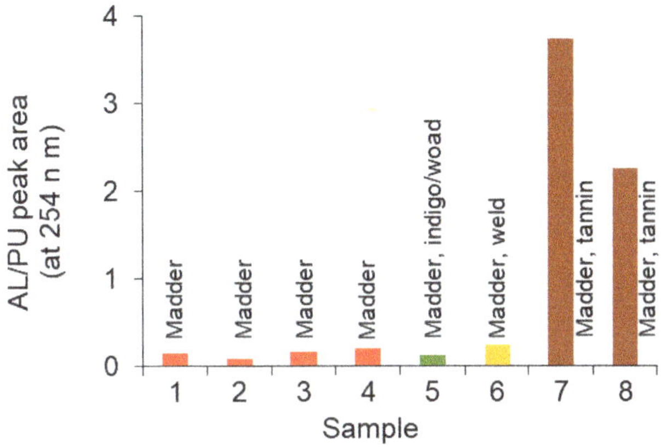

Figure 4. Ratios of HPLC peak areas of alizarin versus purpurin (AL/PU) for textile samples that were dyed with madder. Other dyes were identified in some samples as indicated. Dye identifications were reproduced from Table 2. Peak areas were measured using HPLC chromatograms which were collected at 254 nm.

As shown in Figure 4, a different result was obtained for samples 7 and 8, in which madder was identified along with tannin (ellagic acid). For these samples, the HPLC peak area of alizarin was higher than the corresponding area of purpurin, and the ratio AL/PU was >2.2. Consequently, *R. peregrina* cannot be the biological source of the madder dye used in samples 7 and 8. These samples were most likely dyed with *R. tinctorum; Galium* and *Asperula* species cannot be ruled out.

According to the results of Table 2, indigoid dyes were identified in samples 5 (in mixture with madder), 9–11 and 12 (in mixture with weld). Despite some noteworthy attempts [54], the development of a widely accepted method for distinguishing between indigo (*Indigofera tinctoria* L. and others) and woad (*Isatis tinctoria* L.) by HPLC is not available. Consequently, in Table 2, it is reported that the aforementioned samples could have been dyed either with indigo or woad. Both biological sources contain indigotin and indirubin in lower amounts.

As described in Table 2, in samples 9, 10 and 11, no other dye, apart from indigo/woad, was detected. Samples 10 and 11 were blue in colour whereas sample 9 had a greenish hue (Figure 2 and Supplementary Materials). As described above, in sample 5, indigo/woad and madder were identified. Another mixture often used in antiquity was that of blue indigo/woad with yellow weld to produce green dyed textile fibres [55]. This mixture was used to treat sample 12. The resulting green colour of sample 12 is revealed in the corresponding photograph of Figure 2.

4. Conclusions

The following dyes were identified by HPLC in samples extracted from twelve ancient Egyptian textiles (4th–5th c. AD) of the Museum of Faculty of Archaeology, Cairo University: madder, weld (*Reseda luteola* L.) and indigoid dyes which can be either indigo (*Indigofera* species and other) or woad (*Isatis tinctoria* L.). These identifications were achieved as alizarin, purpurin, rubiadin (madder components), apigenin, luteolin, chrysoeriol (weld components), indigotin and indirubin (indigo/woad components) were detected in the chromatograms of the investigated samples. Moreover, ellagic acid was detected, indicating the presence of tannin sources. In some of the studied samples, madder was the only detected dye. It is reported that these samples were treated with a madder source which was rich in purpurin and poor in alizarin. This result is based on semi-quantitative results obtained from HPLC peak areas, which were measured at 254 nm. In other samples, madder was found in mixture with indigo/woad and weld. The HPLC chromatograms for these samples revealed (again) the presence of a madder source with low alizarin versus purpurin HPLC peak area ratio. These results can offer support to the scenario, which suggests that wild madder (*R. peregrina*) was used to treat the aforementioned samples. However, as alizarin was detected in amounts that are not negligible, the potential use of common madder (*R. tinctorum*) cannot be excluded, particularly if methods were developed by the Egyptian dyers of the past to affect and control the relative composition of madder dye. Madder dyes rich in alizarin (probably originated from *R. tinctorum*) were detected only in samples treated with madder-tannin mixtures. Finally, it is noted that the potential use of other madder species (*Galium* and *Asperula* species) by the Egyptian dyers cannot be ruled out by the chemical results.

Supplementary Materials: The following are available online at https://www.mdpi.com/article/10.3390/heritage4040176/s1, Figure S1. Photographs of samples which were included in the study.

Author Contributions: Conceptualization I.K. and O.A.-K.; methodology, I.K.; investigation, P.K.; data curation, I.K. and D.M.; writing—original draft preparation and review and editing, I.K. and O.A.-K. All authors have read and agreed to the published version of the manuscript.

Funding: This research received no external funding.

Institutional Review Board Statement: Not applicable.

Informed Consent Statement: Not applicable.

Data Availability Statement: Data available upon request from the authors.

Conflicts of Interest: The authors declare no conflict of interest.

References

1. Abdel-Kareem, O. *Principles of Conservation and Restoration of Archaeological Organic Materials, Textiles, Wood, Leather, Manuscripts, Bones, Basketry and Feathers;* Cairo University: Cairo, Egypt, 2015.
2. Wouters, J. Dye Analysis in a Broad Perspective: A Study of 3rd to 10th Century Coptic Textiles from Belgian Private Collections. *Dyes Hist. J. Archaeol.* **1995**, *13*, 38–45.
3. Orska-Gawrys, J.; Surowiec, I.; Kehl, J.; Rejniak, H.; Urbaniak-Walczak, K.; Trojanowicz, M. Identification of Natural Dyes in Archaeological Coptic Textiles by Liquid Chromatography with Diode Array Detection. *J. Chromatogr.* **2003**, *989*, 239–248. [CrossRef]
4. Trojanowicz, M.; Orska-Gawryś, J.; Surowiec, I.; Szostek, B.; Urbaniak-Walczak, K.; Kehl, J.; Wróbel, M. Chromatographic Investigation of Dyes Extracted from Coptic Textiles from the National Museum in Warsaw. *Stud. Conserv.* **2004**, *49*, 115–130. [CrossRef]
5. Verhecken, A. Relation between Age and Dyes of 1st Millennium AD Textiles Found in Egypt. In *Methods of Dating Ancient Textiles of the 1st Millennium AD from Egypt and Neighbouring Countries*; De Moor, A., Fluck, C., Eds.; Lannoo Publishers: Tielt, Belgium, 2007; pp. 206–213.
6. Karapanagiotis, I.; Verhecken-Lammens, C.; Kamaterou, P. Identification of Dyes in Egyptian Textiles of the First Millennium Ad from the Collection Fill-Trevisiol. *Archaeol. Anthr. Sci.* **2019**, *11*, 2699–2710. [CrossRef]
7. Ahmed, H.E.; Tahoun, I.F.; Elkholy, I.; Shehata, A.B.; Ziddan, Y. Identification of Natural Dyes in Rare Coptic Textile Using HPLC-DAD and Mass Spectroscopy in Museum of Faculty of Arts, Alexandria University, Egypt. *Dyes Pigment.* **2017**, *145*, 486–492. [CrossRef]
8. Surowiec, I.; Orska-Gawryś, J.; Biesaga, M.; Trojanowicz, M.; Hutta, M.; Halko, R.; Urbaniak-Walczak, K. Identification of Natural Dyestuff in Archeological Coptic Textiles by HPLC with Fluorescence Detection. *Anal. Lett.* **2003**, *36*, 1211–1229. [CrossRef]
9. Szostek, B.; Orska-Gawrys, J.; Surowiec, I.; Trojanowicz, M. Investigation of Natural Dyes Occurring in Historical Coptic Textiles by High-Performance Liquid Chromatography with UV–Vis and Mass Spectrometric Detection. *J. Chromatogr. A* **2003**, *1012*, 179–192. [CrossRef]
10. Wouters, J.; Vanden Berghe, I.; Richard, G.; Breniauxm, R.; Cardon, D. Dye Analysis of Selected Textiles from Three Roman Sites in the Eastern Desert of Egypt: A Hypothesis on the Dyeing Technology in Roman and Coptic Egypt. *Dyes Hist. Archaeol.* **2008**, *21*, 1–16.
11. Abdel-Kareem, O.; Alawi, M.A.; Mubarak, M.S. Identification of Dyestuffs in a Rare Coptic Garment Using High Performance Liquid Chromatography with Photodiode Array Detection (HPLC-PDA). *J. Text. Appar. Technol. Manag.* **2010**, *6*, 1–7.
12. Karapanagiotis, I.; Theologou, J.; Lakka, A.; Ozoline, A.; Panayiotou, C. Investigation of the Colouring Materials of Fustat Carpet Fragments. *Archaeometry* **2011**, *53*, 587–599. [CrossRef]
13. Gulmini, M.; Idone, A.; Davit, P.; Moi, M.; Carrillo, M.; Ricci, C.; Dal Bello, F.; Borla, M.; Oliva, C.; Greco, C.; et al. The Coptic Textiles of the Museo Egizio in Torino (Italy): A Focus on Dyes through a Multi-Technique Approach. *Archaeol. Anthrop. Sci.* **2017**, *9*, 485–497. [CrossRef]

14. Walton, P. Shellfish Purple in a Coptic Textile. *Dyes Hist. Archaeol.* **1985**, *4*, 33–34.
15. Michel, R.H.; Lazar, J.; McGovern, P.E. Indigoid Dyes in Peruvian and Coptic Textiles of the University Museum of Archaeology and Anthropology. *Archaeomaterials* **1992**, *6*, 69–83.
16. Wouters, J. A New Method for the Analysis of Blue and Purple Dyes in Textiles. *Dyes Hist. Archaeol.* **1992**, *10*, 17–21.
17. Hofmann–de Keizer, R.; Van Bommel, M. Dyestuff Analysis of Two Textile Fragments from Late Antiquity. *Dyes Hist. Archaeol.* **2008**, *21*, 17–25.
18. Liritzis, I.; Laskaris, N.; Vafiadou, A.; Karapanagiotis, I.; Volonakis, P.; Papageorgopoulou, C.; Bratitsi, M. Archaeometry: An Overview. *Sci. Cult.* **2020**, *6*, 49–98.
19. Shahid, M.; Wertz, J.; Degano, I.; Aceto, M.; Khan, M.I.; Quye, A. Analytical Methods for Determination of Anthraquinone Dyes in Historical Textiles: A Review. *Anal. Chim. Acta* **2019**, *1083*, 58–87. [CrossRef] [PubMed]
20. Wouters, J. High Performance Liquid Chromatography of Anthraquinones: Analysis of Plant and Insect Extracts and Dyed Textiles. *Stud. Conserv.* **1985**, *30*, 119–128.
21. Hofenk-de Graaff, J.H. *The Colourful Past: Origins, Chemistry and Identification of Natural Dyestuffs*; Archetype Publications Ltd.: London, UK, 2004.
22. Zhang, X.; Laursen, R.A. Development of Mild Extraction Methods for the Analysis of Natural Dyes in Textiles of Historical Interest Using LC-Diode Array Detector-MS. *Anal. Chem.* **2005**, *77*, 2022–2025. [CrossRef] [PubMed]
23. Valianou, L.; Karapanagiotis, I.; Chryssoulakis, Y. Comparison of Extraction Methods for the Analysis of Natural Dyes in Historical Textiles by High-Performance Liquid Chromatography. *Anal. Bioanal. Chem.* **2009**, *395*, 2175–2189. [CrossRef]
24. Brosseau, C.L.; Gambardella, A.; Casadio, F.; Grzywacz, C.M.; Wouters, J.; Van Duyne, R.P. Ad-hoc Surface-Enhanced Raman Spectroscopy Methodologies for the Detection of Artist Dyestuffs: Thin Layer Chromatography-Surface Enhanced Raman Spectroscopy and in Situ on the Fiber Analysis. *Anal. Chem.* **2009**, *81*, 3056–3062. [CrossRef]
25. Wouters, J.; Grzywacz, C.M.; Claro, A. A Comparative Investigation of Hydrolysis Methods to Analyze Natural Organic Dyes by HPLC–PDA. *Stud. Conserv.* **2011**, *56*, 231–249. [CrossRef]
26. Manhita, A.; Ferreira, T.; Candeias, A.; Dias, C.B. Extracting Natural Dyes from Wool—An Evaluation of Extraction Methods. *Anal. Bioanal. Chem.* **2011**, *400*, 1501–1514. [CrossRef]
27. Santos, R.; Hallett, J.; Conceição Oliveira, M.; Sousa, M.M.; Sarraguça, J.; Simmonds, M.S.J.; Nesbitt, M. HPLC-DAD-MS Analysis of Colorant and Resinous Components of Lac-Dye: A Comparison between *Kerria* and *Paratachardina genera*. *Dyes Pigm.* **2015**, *118*, 129–136. [CrossRef]
28. Van der Klift, E.; Villela, A.; Derksen, G.; Lankhorst, P.; van Beek, T. Microextraction of Reseda luteola-Dyed Wool and Qualitative Analysis of Its Flavones by UHPLC-UV, NMR and MS. *Molecules* **2021**, *26*, 3787. [CrossRef]
29. Marques, R.; Sousa, M.M.; Oliveira, M.C.; Melo, M.J. Characterization of Weld (*Reseda luteola* L.) and Spurge Flax (*Daphne gnidium* L.) by High-Performance Liquid Chromatography–Diode Array Detection–Mass Spectrometry in Arraiolos Historical Textiles. *J. Chromatogr. A* **2009**, *1216*, 1395–1402. [CrossRef]
30. Claro, A.; Melo, M.J.; de Melo, S.S.; Berg, K.J.V.D.; Burnstock, A.; Montague, M.; Newman, R. Identification of Red Colorants in van Gogh Paintings and Ancient Andean Textiles by Microspectrofluorimetry. *J. Cult. Herit.* **2010**, *11*, 27–34. [CrossRef]
31. Han, J.; Wanrooij, J.; van Bommel, M.; Quye, A. Characterisation of Chemical Components for Identifying Historical Chinese Textile Dyes by Ultra High Performance Liquid Chromatography—Photodiode Array—Electrospray Ionisation Mass Spectrometer. *J. Chromatogr. A* **2017**, *1479*, 87–96. [CrossRef]
32. Sharif, S.; Nabais, P.; Melo, M.J.; Oliveira, M.C. Traditional Yellow Dyes Used in the 21st Century in Central Iran: The Knowledge of Master Dyers Revealed by HPLC-DAD and UHPLC-HRMS/MS. *Molecules* **2020**, *25*, 908. [CrossRef]
33. Sanyova, J.; Reisse, J. Development of a Mild Method for the Extraction of Anthraquinones from their Aluminum Complexes in Madder Lakes Prior to HPLC Analysis. *J. Cult. Herit.* **2006**, *7*, 229–235. [CrossRef]
34. Mantzouris, D.; Karapanagiotis, I.; Valianou, L.; Panayiotou, C. HPLC-DAD-MS Analysis of Dyes Identified in Textiles from Mount Athos. *Anal. Bioanal. Chem.* **2011**, *399*, 3065–3079. [CrossRef]
35. Serafini, I.; Lombardi, L.; Vannutelli, G.; Montesano, C.; Sciubba, F.; Guiso, M.; Curini, R.; Bianco, A. How the Extraction Method Could be Crucial in the Characterization of Natural Dyes from Dyed Yarns and Lake Pigments: The Case of American and Armenian Cochineal Dyes, Extracted through the New Ammonia-EDTA Method. *Microchem. J.* **2017**, *134*, 237–245. [CrossRef]
36. Ford, L.; Henderson, R.L.; Rayner, C.M.; Blackburn, R.S. Mild Extraction Methods Using Aqueous Glucose Solution for the Analysis of Natural Dyes in Textile Artefacts Dyed with Dyer's Madder (*Rubia tinctorum* L.). *J. Chromatogr. A* **2017**, *1487*, 36–46. [CrossRef] [PubMed]
37. Kirby, J.; Van Bommel, M.; Coudray, A.; Devièse, T.; Karapanagiotis, I.; Higgitt, C.; Mantzouris, D.; Peggie, D.; Morrison, R.; Gaibor, A.N.P.; et al. From Botanical Source to Analytical Result: The Influence of Recipe and Plant Source on Appearance and Composition of Anthraquinone and Flavonoid Dyes and Pigments. In Proceedings of the 32nd Annual DHA Meeting, Dyes in History and Archaeology (DHA), La Rochelle, France, 3–4 October 2013.
38. Surowiec, I.; Szostek, B.; Trojanowicz, M. HPLC-MS of Anthraquinoids, Flavonoids, and their Degradation Products in Analysis of Natural Dyes in Archeological Objects. *J. Sep. Sci.* **2007**, *30*, 2070–2079. [CrossRef] [PubMed]
39. Karapanagiotis, I.; Mantzouris, D.; Cooksey, C.; Mubarak, M.S.; Tsiamyrtzis, P. An Improved HPLC Method Coupled to PCA for the Identification of Tyrian Purple in Archaeological and Historical Samples. *Microchem. J.* **2013**, *110*, 70–80. [CrossRef]

40. Mantzouris, D.; Karapanagiotis, I.; Panayiotou, C. Comparison of Extraction Methods for the Analysis of Indigofera tinctoria and Carthamus tinctorius in Textiles by High Performance Liquid Chromatography. *Microchem. J.* **2014**, *115*, 78–86. [CrossRef]
41. Karapanagiotis, I. A Review on the Archaeological Chemistry of Shellfish Purple. *Sustainability* **2019**, *11*, 3595. [CrossRef]
42. Karapanagiotis, I.; Lakka, A.; Valianou, L.; Chryssoulakis, Y. High-Performance Liquid Chromatographic Determination of Colouring Matters in Historical Garments from the Holy Mountain of Athos. *Microchim. Acta* **2008**, *160*, 477–483. [CrossRef]
43. Cardon, D. *Natural Dyes—Sources, Tradition, Technology and Science*; Archetype Publications Ltd.: London, UK, 2007; p. 107.
44. Shahid, M.; Ahmad, A.; Yusuf, M.; Khan, M.I.; Khan, S.A.; Manzoor, N.; Mohammad, F. Dyeing, Fastness and Antimicrobial Properties of Woolen Yarns Dyed with Gallnut (*Quercus infectoria* Oliv.). *Extract. Dyes Pigm.* **2012**, *95*, 53–61. [CrossRef]
45. Newman, R.; Gates, G.A. The Matter of Madder in the Ancient World. In *Mummy Portraits of Roman Egypt: Emerging Research from the APPEAR Project*; Svoboda, M., Cartwright, C.R., Eds.; J. Paul Getty Museum: Los Angeles, CA, USA, 2020; pp. 24–33.
46. Plants of de World Online. Available online: http://www.plantsoftheworldonline.org (accessed on 29 September 2021).
47. Karapanagiotis, I.; Minopoulou, E.; Valianou, L.; Daniilia, S.; Chryssoulakis, Y. Investigation of the Colourants Used in Icons of the Cretan School of Iconography. *Anal. Chim. Acta* **2009**, *647*, 231–242.
48. Petroviciu, I.; Cretu, I.; Vanden Berghe, I.; Wouters, J.; Medvedovici, A.; Albu, F.; Creanga, D. A Discussion on the Red Anthraquinone Dyes Detected in Historic Textiles from Romanian Collections. *e-PS* **2012**, *9*, 90–96.
49. Mouri, C.; Laursen, R. Identification of Anthraquinone Markers for Distinguishing Rubia Species in Madder-Dyed Textiles by HPLC. *Microchim. Acta* **2012**, *179*, 105–113. [CrossRef]
50. Mantzouris., D.; Karapanagiotis, I. Armenian Cochineal (*Porphyrophora hamelii*) and Purpurin-Rich Madder in Ancient Polychromy. *Color. Technol.* **2015**, *131*, 370–373. [CrossRef]
51. Sukenik, N.; Iluz, D.; Amar, Z.; Varvak, A.; Workman, V.; Shamir, O.; Ben-Yosef, E. Early Evidence (Late 2nd Millennium BCE) of Plant-Based Dyeing of Textiles from Timna, Israel. *PLoS ONE* **2017**, *12*, 0179014. [CrossRef] [PubMed]
52. Wouters, J. The Dye of Rubia peregrina—Preliminary Investigations. *Dye. Hist. Archaeol.* **2001**, *16/17*, 145–157.
53. Daniels, V.; Devièse, T.; Hacke, M.; Higgitt, C. Technological Insights into Madder Pigment Production in Antiquity. *Tech. Res. Bull.* **2014**, *8*, 13–28.
54. Witkowski, B.; Ganeczko, M.; Hryszko, H.; Stachurska, M.; Gierczak, T.; Biesaga, M. Identification of Orcein and Selected Natural Dyes in 14th and 15th Century Liturgical Paraments with High-Performance Liquid Chromatography Coupled to the Electrospray Ionization Tandem Mass Spectrometry (HPLC-ESI/MS/MS). *Microchem. J.* **2017**, *133*, 370–379. [CrossRef]
55. Karapanagiotis, I.; Karadag, R. Dyes in Post-Byzantine and Ottoman Textiles: A Comparative Study. *Mediterr. Archaeol. Archaeom.* **2015**, *15*, 177–189.

heritage

MDPI

Article

Experimental Investigations and Microstructural Characterization of Construction Materials of Historic Multi-Leaf Stone-Masonry Walls

Osama Amer [1,*], Danila Aita [2], Ezzeldin K. Mohamed [3], Akram Torky [4] and Ashraf Shawky [4]

1 Department of Conservation, Cairo University, Giza, Cairo 12613, Egypt
2 Architecture and Design Department, University of Genoa, 16121 Genoa, Italy; danila.aita@edu.unige.it
3 National Water Research Centre, Construction Research Institute, Shoubra El-Kheima 13621, Egypt; ee.k.mohamed@gmail.com
4 Department of Civil Engineering, Faculty of Engineering, Cairo University, Giza, Cairo 12613, Egypt; aktorkey@yahoo.com (A.T.); ashraf.shawky63@gmail.com (A.S.)
* Correspondence: osama_amer@cu.edu.eg

Abstract: In order to correctly define the pathology of multiple-leaf stonemasonry walls and determine the appropriate interventions for its conservation and preservation, comprehensive studies on its building materials should be carried out since the overall behaviour of masonry structures is highly dependent on the characterization of its construction materials. Consequently, an interdisciplinary procedure for construction material characterization used in multiple-leaf stone-masonry walls in Egypt has been implemented to enrich documentation, conservation and restoration issues of this type of wall. The research methodology integrates experimental data obtained through on-site sampling, conducted tests and analyses, historical information, and field survey observations. The fundamental physical and mechanical properties of the masonry elements were examined by incorporating stone blocks, mortars and core-infill materials. The mineralogical composition and interlocking textures of the collected samples were investigated utilizing a large range of complementary investigation and analysis techniques, including polarizing microscopy, X-ray diffraction (XRD), thermal analysis (TG/DTA), and environmental scanning electron microscope (ESEM) attached to an EDX unit. Through the results thus obtained, a complete characterization of the mineralogical composition; physical–mechanical, chemical, and thermal properties; and the interlocking textures of the construction materials of both the outer and inner-core layers was performed. The outer leaves of the majority of the multiple-leaf stone-masonry walls in medieval architectural heritage were mainly built of well-dressed limestone blocks with nearly uniform dimensions, while the inner-core layer was usually built of stone-rubble infill with bending lime-based mortar. The uniaxial compressive strengths of core infill (corresponding to the inner core layer) and lime-based mortar of the embedded joints are shown to be 85 and 92.5% lower than the limestone units of the outer layer, respectively. Moreover, experimental observations indicate that the inner core layer exhibits the highest porosity values; consequently, deteriorated, loose and cohesionless core infill could greatly affect the durability and thermal resistivity of this kind of wall. The results provide scientific support for investigating the overall structural behaviour of this type of walls and for decision-making in future conservation and restoration strategies.

Keywords: masonry walls; construction materials; architectural heritage; microanalysis; mineralogical investigation

Citation: Amer, O.; Aita, D.; Mohamed, E.K.; Torky, A.; Shawky, A. Experimental Investigations and Microstructural Characterization of Construction Materials of Historic Multi-Leaf Stone-Masonry Walls. *Heritage* **2021**, *4*, 2390–2415. https://doi.org/10.3390/heritage4030135

Academic Editor: Nikolaos Lazaridis

Received: 1 August 2021
Accepted: 7 September 2021
Published: 14 September 2021

1. Introduction

Historical buildings are considered an expression of the science, culture and history of their builders. Masonry has been used to construct the most long-lasting ancient monuments and is present in the most impressive historical structures as evidence of the spirit of enterprise of ancient cultures [1–3]. It is a traditional, widely used, extremely flexible and economical construction method with considerable potential for future development [4,5]. Moreover, masonry has been used to construct significant structures since the beginning of civilization for its robustness, durability and for aesthetic reasons [6].

Generally, masonry is the generic term for a composite material made of units that are usually laid in and solidly bound together, employing mortar or just friction forces between the blocks in many different arrangements. It is a heterogeneous material whose components present a relatively unknown geometry and a high mechanical variability [7]. Therefore, characterizing its construction materials is a topic of great interest as regards the accurate design of the most proper intervention techniques and materials.

The present research deals with a distinctive type of historical masonry wall, namely multiple-leaf stone-masonry (MLSM) walls (Figure 1). Most of the medieval complex historic constructions present structural elements built by adopting multiple-leaf masonry technology. This type of masonry wall is common in historic Islamic architectural heritage in Egypt, as well as in worldwide monumental structures, which identifies a non-homogeneous type of structural element.

Figure 1. Typical examples of multiple-leaf stone masonry walls: (**a**,**b**) *Complex of Sultan al-Ashraf Barsbay* (835 A.H.–1433 A.D.); (**c**,**d**) *Zawiya of 'Ali al-Maghrabi* (1282 A.H.–1866 A.D.); (**e**) *Complex of Al-Sultan Hasan (757–764 A.H.–1356–1362 A.D.)*; (**f**) *Qasr al-Amir Qawsun* (738 A.H.–1337 A.D.); (**g**) *Saray Al-Musafir khana* (1193–1203 A.H.–1779–1788 A.D.).

This building technique was used for vertical structural elements in historic houses, public buildings, religious constructions, and also for piers in roads and railway bridges.

The actual mechanical behaviour of such structural elements is undoubtedly affected by the low resistance in traction of the constituents, particularly as regards the weaker component. The behaviour of multi-leaf masonry walls has been reported in the literature [8–14], as well as material characterization of three-leaf masonry walls [3,15–18] and the utilized strengthening techniques [19–24]. The characterization and assessment of the degradation state of historical buildings, including multi-leaf masonry walls, have also been reported in the literature [3,25,26].

Through on-site observations and measurements, along with laboratory tests, a systematic survey was conducted concerning the evolution of the constructive technique and building materials utilized for multiple-leaf masonry load-bearing walls by considering historic buildings in Egypt corresponding to different eras. Different samples of the building materials used in the construction of multiple-leaf masonry walls were collected from various medieval historic buildings in Cairo, including stone blocks, mortar from bed and head joints, and filling materials. The survey's objective was to diversify the buildings from which the samples were derived to represent the building materials used in these types of walls throughout different eras. The current experimental campaign is devoted to characterizing the construction materials used in multiple-leaf stone-masonry walls. These analyses and examinations were conducted on the masonry composing elements (stone, natural lime mortar, and core infill materials); they were aimed at accurately characterizing the main chemical, mineralogical and thermal properties of construction materials and correctly defining the pathology of multiple-leaf masonry walls in order to determine the most appropriate intervention techniques.

2. Materials and Methods

The research methodology followed in this study is divided into five tasks: (I) Mineralogical analyses through qualitative analysis using XRD and thermal analysis. These analyses are conducted for various specimens of limestone and lime-based mortars collected, during the field survey, from selected historic multiple-leaf masonry walls to determine their chemical composition and constituent minerals and compounds. (II) Petrographic investigation and thin-section analyses using polarizing microscopes in order to investigate the mineralogical composition and interlocking textures of the construction materials under examination (i.e., limestone and lime-based mortars). (III) Identification of the microstructure and morphological examinations of multiple-leaf masonry wall construction materials using environmental scanning electron microscopy (ESEM) with qualitative analysis using EDAX. (IV) Characterization of the masonry composing elements through physical and mechanical tests. (V) Thermal properties identifying the thermal behaviour, resistance and conductivity of such materials.

2.1. Field Survey

The study of the multiple-leaf masonry wall cross-sections has various aims concerning diagnosis and conservation work. One of the most significant objectives of the present field survey is to provide essential input data concerning the multiple-leaf masonry wall morphology and geometry to provide a reliable scientifically-based structural assessment that allows the formulation of valid hypotheses on its mechanical behaviour and failure mechanisms. To achieve these goals, thirty-three archaeological Islamic buildings in Cairo dating to different Islamic eras and at various locations were surveyed. Various recorded stone blocks and mortar samples were studied, while the research sources provided histories and previous documentation of some of the surveyed buildings. The conducted field survey focused on the construction technology of multiple-leaf stone-masonry walls, building materials, the overall thickness of the walls, the ratio between inner and external leaves, the composition of the inner core layer and the connectivity between the walls' leaves. For further details of the field survey, the interested reader may refer to [3,27].

2.2. Sampling

The testing program includes extracting a set of 55 representative limestone, mortar and core-infill samples from the investigated multiple-leaf stone-masonry wall of various historic buildings in Egypt. Thirty stone samples were derived from the outer and inner layers, and fourteen mortar samples were derived from the bed and head joints of the outer layer, while eleven core-infill samples were derived from the inner core layer. All the samples here analysed and investigated were carefully collected from areas without aesthetic value or from severely damaged parts.

2.3. Petrographic Investigation

The petrographic investigation of stone and mortar samples includes determining the mineral content, grain size, micro fracturing and interlocking texture. Petrographic microscopy investigation was conducted using a Polarizing petrographic microscope on thin sections of the stone and mortar samples. Air-dried samples (at ~40 °C to avoid dehydration of components, especially Gypsum if present, and physical damage due to thermal shock) were subjected to impregnation with warmed low viscosity colour dyed epoxy resin, to aid in the visualization of pores, cracks and air voids. Very thin slices of collected samples (cut perpendicular to the bedding planes) were mounted on clear, flat glass slides. The thickness reduction (to 20–30/am) permits light to pass through crystalline or amorphous materials; this is an important aspect for the detailed analysis and recognition of the stone and mortar's components. Morphological examination of the prepared thin sections of the stone and mortar samples was carried out using a NIKON OPTI PHOTO x23 polarized transmitted light microscope equipped with photo camera S23 under cross-polarized light XPL.

2.4. Mineralogical Characterization

Construction materials of multiple-leaf stone-masonry walls were studied by laboratory investigation. The samples collected during the field survey that were derived either from the outer leaves or inner-core layer of different historic Islamic constructions were studied as regards their mineralogical composition, which was identified by X-ray diffraction and thermal analysis as described in the following sections.

2.4.1. Quantitative Analysis Using X-ray Diffraction Technique (XRD)

The basic principles of this technique have been described in detail by [28–32], among others.

Mineral compositions of the collected samples were identified by means of X-ray diffraction patterns, using an *X′Pert PR* PAN analytical X-ray diffraction model with a secondary monochromator. The samples under examination were prepared for the analysis by drying at 110 °C and grinding to less than 75 mm diameter. The analysis was run using Ni-filter and Cu-Kα radiation (λ = 1.542 Å) at 45 kV, 35 MA and scanning speed 0.02° (2θ)/sec. The diffraction peaks between 2θ = 0° and 60° and corresponding spacing (d, Å) and relative intensities (I/I°) were obtained. The diffraction charts and relative intensities were obtained and compared with ICDD files.

2.4.2. Thermal Analysis (TG/DTA)

As corroborative tools, thermogravimetric analyses/differential thermal analyses (TGA/DTA) were carried out on the stone and mortar samples collected from the different multiple-leaf stone-masonry walls. The thermal analysis involves studying the evolution of several physical properties as a function of temperature. When the material is subjected to heating or cooling, its chemical composition and crystal structure undergo various changes such as reaction, oxidation, decomposition, fusion, expansion, contraction, crystallization, or phase transition. All these changes can be detected using differential thermal analysis [33]. Furthermore, thermal transformations like dehydration, dihydroxylation, oxidation, and decomposition can be revealed through thermal investigations [34].

Thermogravimetric analysis (TGA) deals with the change in the mass of a substance continuously monitored as a function of temperature or time when it is heated or cooled at a predetermined rate. It provides information on the thermal stability of the sample at different temperatures, the purity of the sample, as well as its water, carbonate, and organic content. It is also helpful for studying decomposition reactions [33]. Differential thermal and thermogravimetric analyses (TGA/DTA) are suitable to establish characteristics of the stone and mortar samples, as it is easy to detect the main components, the nature of the aggregate, and other aspects with a small quantity of sample [34–36] and to define the temperature at which material starts decomposing [37]. Moreover, it is also possible to know whether the decomposition occurs in one or more stages by observing the non-linear drops in the sample's mass [33,38].

Thermal analysis experiments were performed using thermogravimetry and differential thermal analysis (DT.50 thermal analyser, Shimadzu Co., Kyoto, Japan). The stone and mortar samples to be analysed by TGA/DTA were dried and ground up to 106 μm. The experiment was carried out by heating the sample from room temperature up to 1000 °C at a rate of 10 °C/min in a static air atmosphere. The sensitivity of this equipment is 0.0001 mg with a temperature range from ambient temperature to 1000 °C.

2.5. Microstructure and Micro-Morphological Examination Using SEM with EDX

A scanning electron microscope (SEM) with an EDX (microanalysis) attachment was used to identify the structural morphology and microstructure of the collected stone and mortar samples, determine the forming minerals, and observe voids and weathering status. Energy-dispersive X-ray analysis (EDX) was used to determine the elemental composition where necessary, which proved to be complementary information to support the previous techniques employed. Qualitative analysis using EDX was performed on uncoated samples to avoid overlapping gold peaks with beaks of interest, using a higher accelerating voltage of 20 kV and a larger spot size than the imaging. EDX spectra were obtained between 0 and 10 KeV.

SEM-EDX analysis was carried out using an SEM Model Quanta 250 FEG (Field Emission Gun) attached with an EDX unit (energy dispersive X-ray analysis), with an accelerating voltage of 30 K.V, magnification of 14× up to 1,000,000×, and resolution for Gun.1n. EDX spectra were recorded in the spot-profile mode by focusing the electron beam onto specific regions of the sample. The following is a detailed SEM examination and EDX micro-analysis of the collected stone and mortar samples under investigation.

2.6. Physical Characterization

According to the previous microanalysis on the mineralogical and chemical composition of the elements of MLSM walls [3], the mortar and core-infill specimens were selected to correctly represent the physical properties of the historic construction materials. The specimens used for the physical tests were 40 ± 5 mm cubes; a total of 18 specimens were tested; six specimens were considered for each type of masonry constituent (stone, mortar and core-infill). The core infill specimens had a percentage of stone fragments of about 60–70% in volume, in accordance with the common cases observed in the conducted field survey. The physical properties included porosity (η), dry density (ρ_d), bulk density (ρ_{Bulk}), water absorption (WA), specific weight (Gs) and void ratio (e), which were determined following the procedures outlined in ISRM's suggested methods [39].

2.7. Mechanical Characterization

In the preparation of the specimens, special care was taken to ensure parallel ends with a smooth surface and perpendicular to the longitudinal axis of the tested specimens. All specimens were oven-dried and tested in environmental laboratory conditions. The experimental campaign on the composing elements (stone, lime-based mortar and core-infill) focused on determining the uniaxial compressive and tensile strength, in accordance with the ASTM D2938 standard [40] and the suggested testing methods by [41,42], a uniaxial testing machine with a hydraulic actuator was used. The tests were carried out under displacement control at a displacement rate of 10μm/s, permitting the tracing of the softening path to avoid the explosive failures originated by the axial load control. The complete stress–strain diagrams that characterize the compressive behaviour of all specimens are the result of averaging the displacement recorded by the LVDT between the upper and lower steel plates of the testing machine.

Regarding the lime-based mortar and core-infill, the specimens were left under controlled conditions for approximately 120 days. The same lime-based mortar was used as a binder in core-infill specimens with limestone fragments as a filler. The average size of used limestone fragments is between 1 and 10 cm in diameter.

2.8. Thermal Conductivity and Resistivity

The thermal conductivity of the lime-based mortar, limestone and core infill was evaluated utilizing a heat flow meter apparatus according to [43–45]. All the specimens were assembled with external dimensions 300 × 300 mm due to the dimensions of the experimental apparatus (Figure 2).

Figure 2. Tested specimens by means of the hot plate for thermal measurements (**a**) stone, (**b**) lime-based mortar, and (**c**) core-infill samples; (**d**) heat flow meter apparatus.

Following the previously conducted microanalysis on the mineralogical and chemical composition of the stone, mortar and core-infill samples, the specimens under examination were selected to correctly represent the actual properties of the historic construction materials used in multiple-leaf masonry walls.

The core infill specimens had a percentage of stone fragments of about 60–70% in volume. The thermal behaviour of stone (S_1, S_2), mortar (M_1, M_2) and core-infill specimens (RM_1, RM_2) were investigated with a laser comp instrument (*Thermo-CUBE*) with a self-contained water reservoir and a mean specimen temperature range of 5 °C to 40 °C. The steady-state method was used, applying Fourier's law of heat conduction to measure thermal conductivity. The specimens were interposed between two parallel plates (the hot guard heater and the cold plate) at constant but different temperatures. The power rate input in the hot plate with the metered area, A, is measured when thermal equilibrium is reached at steady-state conditions. When the control system is used, the plate temperatures reach stability [46]. The tested specimens with a thickness of 40 mm were placed between two flat plates controlled to a uniform one-dimensional temperature field. The temperature drop across the specimen is measured by thermocouples fixed in the plates, while the heat flow through the specimen is measured by wireless thermal flux meters (HFMs) embedded in the centre of each plate. Figure 3 illustrates a schematic diagram of the heat flow meter apparatus.

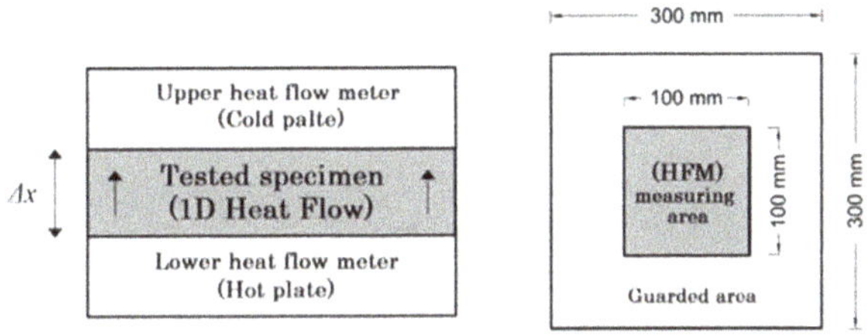

Figure 3. Schematic diagram of the Fox 314 heat flow meter apparatus and plan view of the upper/lower plate showing the guarded area. Adapted from [47].

After thermal equilibrium was developed, where the heating/cooling plates were kept at stable temperatures, the thermal conductivity (λ, in W/M·K) was calculated by measuring the heat flux (Φ, in W/m^2), the temperature difference across the specimen (ΔT, in °C) and the thickness of the specimen (Δx, in m) using the following unidirectional steady-state heat transfer equation (Equation (1)).

$$\lambda_{eff} = \frac{\Phi \, \Delta x}{\Delta T} \tag{1}$$

where Φ is the heat flux (W/m^2) flowing through the specimen, λ_{eff} is the thermal conductivity (W m^{-1} K^{-1}) of the specimen, Δx is the sample thickness (m), and ΔT is the difference in temperature between the hot and cold surfaces of the specimen (°C) ($\Delta T = T_{hot} - T_{cold}$).

3. Results and Discussion

3.1. Characterization of the Cross-Section Morphology

Based on the collected data and field survey outcomes, most of the thick walls of the historic Islamic period were built of multiple-leaf masonry systems at the lower floor levels to acquire thicker wall sections and consequently higher weights (that increases the overall stability of the wall) at a relatively low economic cost. Moreover, the majority of the multiple leaf masonry walls had very thick sections (ranges from 0.5 to more than 6 m) with a homogeneous distribution of the regular well-dressed stone blocks with nearly constant or uniform dimensions in the external leaves (ashlars facing system). Much less homogeneous distribution of the stone-rubble units (as they were randomly cut from quarries and adhered with mortar) was often used as an infilling between the joints among the undressed and rough stone-rubble with variable joint thicknesses.

Furthermore, the external layers were usually built of well-dressed limestone blocks with nearly uniform dimensions (ashlar limestone facing system), while the dimensions of the stone-rubble infill vary according to the total thickness of the inner layer. Moreover, the joint thickness was usually much lower than the thickness of stone-rubble units in the inner core layer, and the percentage of voids and mortar in the inner-core layer is relatively low; consequently, rubble-stones settle at a high percentage. In many cases, a transversal connection between the inner and outer layers is provided by irregularly passing elements. These transversal bond elements were used to enhance the connectivity between the wall's leaves (see Figure 4).

Figure 4. Transversal bond elements to enhance the connectivity between wall leaves: (**A**) external wall of *Bab Zuwayla* and (**B**) main façade of *Masjid al-Salih Tala'i'*.

Moreover, adding through stones to the wall's cross-section improves the deformation capacity, mechanical performance, and out-of-plane strength. The contribution of the transversal bond element (also referred to as through stone) to enhancing the seismic performance and capacity of unreinforced masonry walls has been studied in the literature from the experimental and computational modelling point of view [48–50].

The survey results with the collected and analysed data made it possible to identify the core-infill materials of the multiple-leaf masonry walls. The core layer of the majority of these walls are built mainly of stone-rubble infill. Nevertheless, the core layer in some buildings dating back to the Ottoman period was mainly built of brick-rubble, such as *Sabil Ibrahim Agha Mustahfizan* and *Saray Al-Musafir khana* (Figure 5). Moreover, the thickness of the core layer of the majority of this kind of wall ranges from 0.5 to 2 m, and the percentage of the core layer with thickness lower than 0.5 m is about 14.5%.

Figure 5. Different core-infill materials of multiple-leaf masonry walls: (**A–C**) stone-rubble infill, (**D**,**E**) brick-rubble infill, (**A**,**C**) *Complex of Sultan al-Ashraf Barsbay*, (**B**) *Khanqat Khawand Umm Anuk*, (**D**) *Sabil Ibrahim Agha Mustahfizan*, (**E**) *Saray Al-Musafir khana Zawiya of 'All al-Maghrabi*.

The typology of the surveyed multiple-leaf masonry walls could be characterized from the analysis of collected data and field survey results. The most frequent stone masonry typology comprises three leaves not interconnected, with the outer leaves made of well-dressed limestone blocks with nearly uniform dimensions (ashlars limestone facing system) bonded in horizontal courses. The average thickness of the walls' cross-section is about 1 to 2 m with a thickness ratio between the external and the internal leaves of about 0.15 to 0.3 (Figure 6). The average frequency of voids percentage on core skin is around 2 to 15%. A general characterization of the common topologies of the transversal section of multiple-leaf stone-masonry walls is described in Figure 7.

Figure 6. Various typologies of multiple-leaf stone-masonry walls in historic Islamic buildings: (**A**,**B**) slender wall (*Complex of Sultan al-Ashraf Barsbay* and *Bab Qaytbay resbectivly*), (**C**,**D**) wall of medium thickness (*Complex of Sultan al-Ashraf Barsbay*), (**E**) thick wall, (Zawiya of 'All al-Maghrabi), (**F**) monolithic/massive wall (*Bab Qaytbay*).

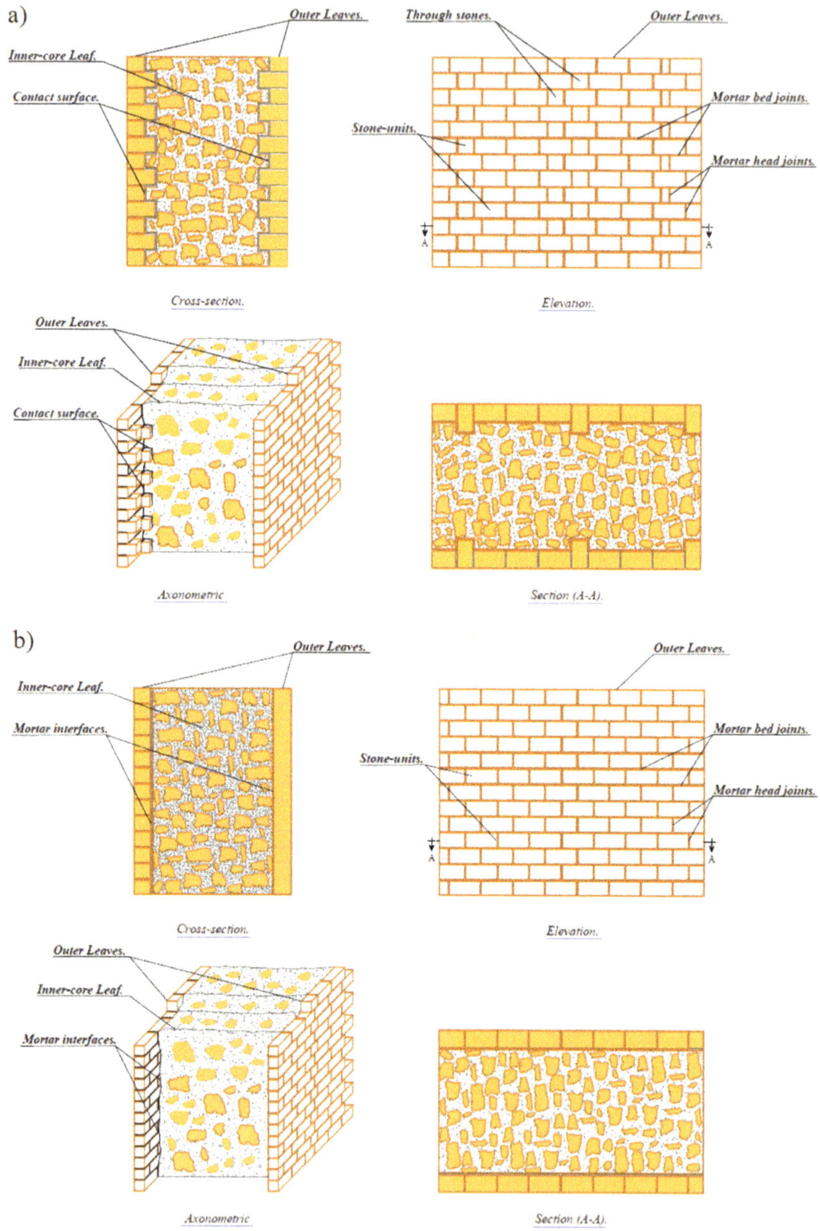

Figure 7. Common topologies of the transversal section of multiple-leaf stone-masonry walls in historic mediaeval masonry constructions in Egypt (**a**) with keyed and (**b**) straight collar joints.

3.2. Petrographic Investigation

The interlocking texture and microstructure of the studied stone samples are given in Figure 8a under cross-polarized light (XPL). These samples represent limestone units derived from the external and inner layers of the investigated multiple-leaf rubble stone-masonry walls. Most of the stone units are fine-grained and composed essentially of carbonate minerals, calcite, a minor amount of dolomite, and a small amount of quartz, phosphate mineral and iron oxides. Carbonate minerals appear as very fine-grained and constitute the matrix of the stone enclosing other constituents. Many microfossils and shells of different shapes and sizes are scattered in the very fine-grained matrix of carbonate minerals and filled with recrystallized carbonate minerals. Quartz grains are very fine-grained and scattered in the stone. Some parts of the sample are slightly stained by iron oxides.

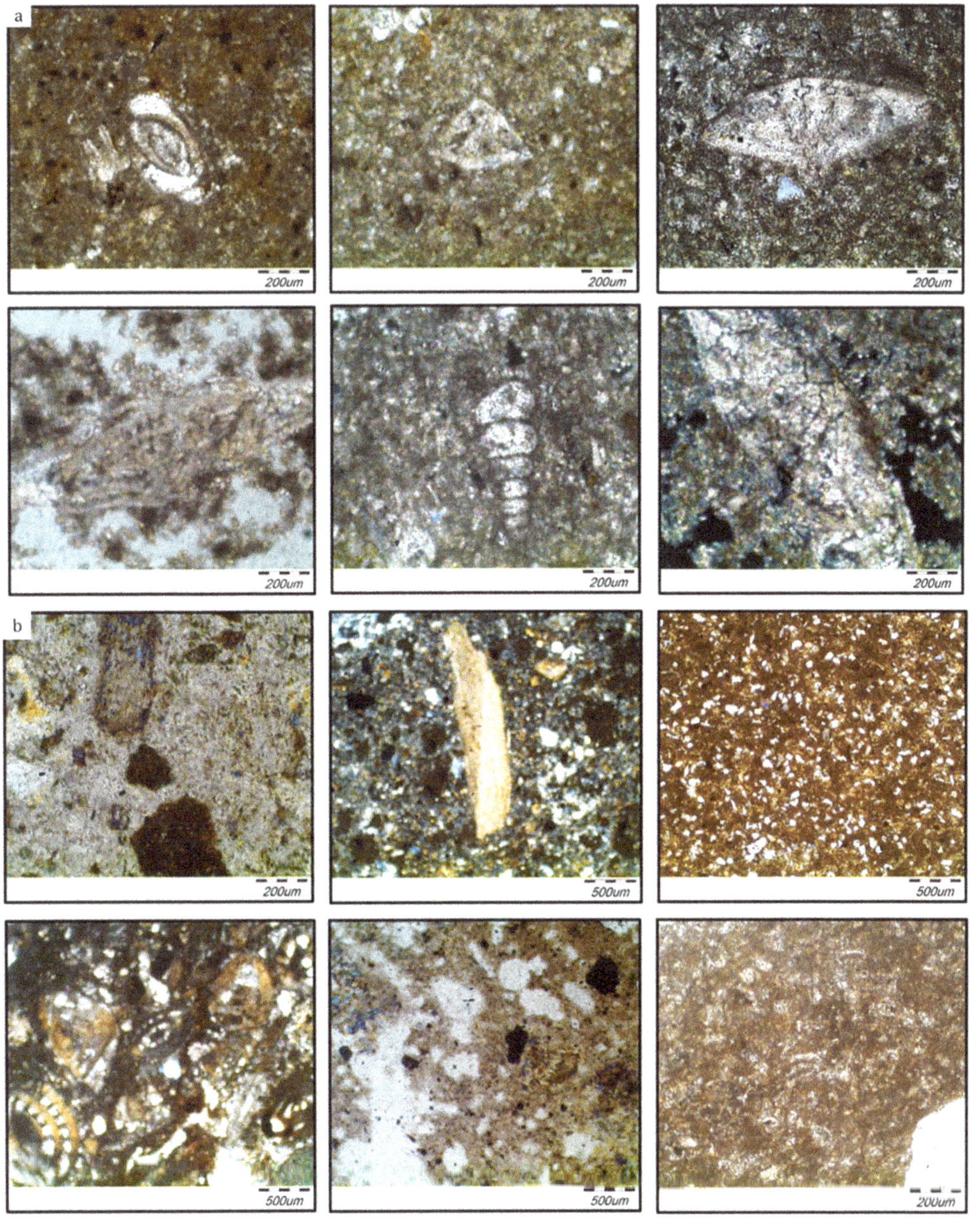

Figure 8. Thin section micrographs of (**a**) limestone samples; (**b**) mortar samples under cross polarized light (XPL).

On the other hand, the interlocking texture and microstructure of the mortar samples examined here are given in Figure 8b under cross-polarized light (XPL). These samples represent mortar derived from the bed joints of the external and also from the inner layers of the investigated walls. By studying the interlocking texture of the mortar samples, it was confirmed that the mortar grains are very fine to medium-grained and composed of either only carbonate minerals or carbonate minerals and gypsum as the major components with minor amounts of quartz and clay minerals and a small amount of mica, iron oxides and opaques. Carbonate minerals occur as very fine-grained aggregates and represent the essential constituent of the mortar admixed with gypsum, while gypsum occurs as very fine to fine-grained, fibrous aggregates and is admixed with carbonates. Quartz, mica, iron oxides and opaques are scattered in a very fine matrix of carbonates. Quartz occurs as fine-grained, sub-rounded to subangular in shape and sometimes encloses mica. Opaques appear as fine-grained and scattered around the sample.

3.3. Quantitative Analysis Using X-ray Diffraction Technique (XRD)

The outer leaves of most of the surveyed multiple-leaf stone-masonry walls were mainly built of limestone blocks. In addition, the core-infill layer is built of rubble with bending mortar; this rubble commonly consists of rough and undressed limestone. According to the XRD analysis results, limestone is mainly composed of either calcite $CaCO_3$ or calcite-magnesian $(Mg_{0.064}\ Ca_{0.936})\ (CO_3)$. Sometimes gypsum $CaSO_4\ (H_2O)_2$ is found at a low percentage as a result of salt decay.

According to [51], gypsum is created from the reaction of calcite ($CaCO_3$) with sulphuric acid (H_2SO_4) in intensively decaying conditions under the influence of natural and anthropogenic factors such as atmospheric humidity and temperature changes, air pollution, salts, and aggressive microbial communities. This gypsum-rich patina develops as black crusts of various thicknesses and extensions. Quartz is also found at a meagre percentage (i.e., traces) as an impurity. Table 1 summarizes the average composition of the minerals for limestone samples collected from different multiple-leaf stone-masonry walls of historic Islamic buildings in Cairo.

Table 1. Average composition of the minerals for limestone samples.

Sample No.	Minerals	Chemical Formula	Semi-Quant [%]
1	Calcite	$CaCO_3$	79
	Quartz	SiO_2	21
2	Calcite, magnesian	$(Mg_{0.064}\ Ca_{0.936})\ (CO_3)$	71
	Gypsum	$CaSO_4\ (H_2O)_2$	23
	Barite	$BaSO_4$	6
3	Calcite, magnesian	$(Mg_{0.064}\ Ca_{0.936})\ (CO_3)$	74
	Gypsum	$CaSO_4\ (H_2O)_2$	18
	Quartz	SiO_2	8
4	Calcite, magnesian	$(Mg_{0.064}\ Ca_{0.936})\ (CO_3)$	100
5	Calcite, magnesian	$(Mg_{0.064}\ Ca_{0.936})\ (CO_3)$	85
	Quartz, low, syn	SiO_2	5
	Halite	$NaCl$	10
6	Calcite, magnesian	$(Mg_{0.064}\ Ca_{0.936})\ (CO_3)$	77
	Gypsum	$CaSO_4\ (H_2O)_2$	23
7	Calcite, magnesian	$(Mg_{0.064}\ Ca_{0.936})\ (CO_3)$	71
	Gypsum	$CaSO_4\ (H_2O)_2$	18
	Quartz	SiO_2	11
8	Calcite	$CaCO_3$	76
	Quartz, low, syn	SiO_2	15
	Halite	$NaCl$	9

Regarding the analysed mortar samples, the analysis results confirmed that lime-based mortar is the most common type of mortar used in the construction of both inner-core layer and external layers as a primary binder between stone-blocks, in the case of external layers, or between rubble-stones, in the case of the inner-core layer.

In accordance with the XRD analysis results, mortar samples, collected either from the inner or external layers, are mainly composed of lime as the major binders, with sand as an aggregate and some additives used to enhance the adhesion performance of the mortar, such as red-brick powder (i.e., *hommra*) or fly ash (i.e., *qusrmil*) as pozzolanic materials. Sometimes gypsum $CaSO_4\ (H_2O)_2$ is found in varying percentages.

The presence of gypsum in the composition of the mortar can be interpreted either as an alternative binder to lime, as an additive to lime mortar mixture enhancing the setting and hardening processes or as a result of salt decay from the reaction of calcite ($CaCO_3$) with sulphuric acid (H_2SO_4) in intensively decaying conditions under the influence of natural and anthropogenic factors. Table 2 summarizes the average composition of the mortar samples collected from different multiple-leaf stone-masonry walls of historic Islamic buildings in Cairo. XRD charts related to the output results of representative stone and mortar samples are given in Figures 9 and 10.

Table 2. Average composition of the minerals for the representative mortar samples.

Sample No.	Minerals	Chemical Formula	Semi-Quant [%]
1	Calcite, magnesian	$(Mg_{0.064}\ Ca_{0.936})\ (CO_3)$	39
	Quartz	SiO_2	19
	Vermiculite-2M	$Mg_3Si_4O_{10}\ (OH)_2$	24
	Kaolinite-1A	$Al_2Si_2O_5\ (OH)_4$	18
2	Gypsum	$CaSO_4\ (H_2O)_2$	58
	Calcite	$CaCO_3$	24
	Quartz	SiO_2	6
	Anhydrite	$CaSO_4.1/2\ H_2O$	12
3	Calcite, magnesian	$(Mg_{0.064}\ Ca_{0.936})\ (CO_3)$	54
	Quartz, syn	SiO_2	19
	Albite, calcian, ordered	$(Ca, Na)\ Al\ (Al, Si)_3\ O_8$	23
	Gypsum	$CaSO_4\ (H_2O)_2$	4
4	Quartz	SiO_2	92
	Calcite	$CaCO_3$	8
5	Calcite	$CaCO_3$	9
	Gypsum	$CaSO_4\ (H_2O)_2$	91
6	Calcite, magnesium, syn	$(Mg_{0.064}\ Ca_{0.936})\ (CO_3)$	51
	Quartz, syn	SiO_2	38
	Hematite, syn	Fe_2O_3	11
7	Halite	$NaCl$	31
	Quartz	SiO_2	56
	Calcite, magnesian	$(Mg_{0.064}\ Ca_{0.936})\ (CO_3)$	13
8	Dolomite	$CaMg\ (CO_3)_2$	12
	Albite	$NaAlSi_3O_8$	33
	Hollandite (Ti, Mg)	$Ba_{6.00}\ Ti_{34.00}\ Mg_{6.00}\ O_{80.00}$	27
	Calcite	$CaCO_3$	8
	Quartz	SiO_2	20

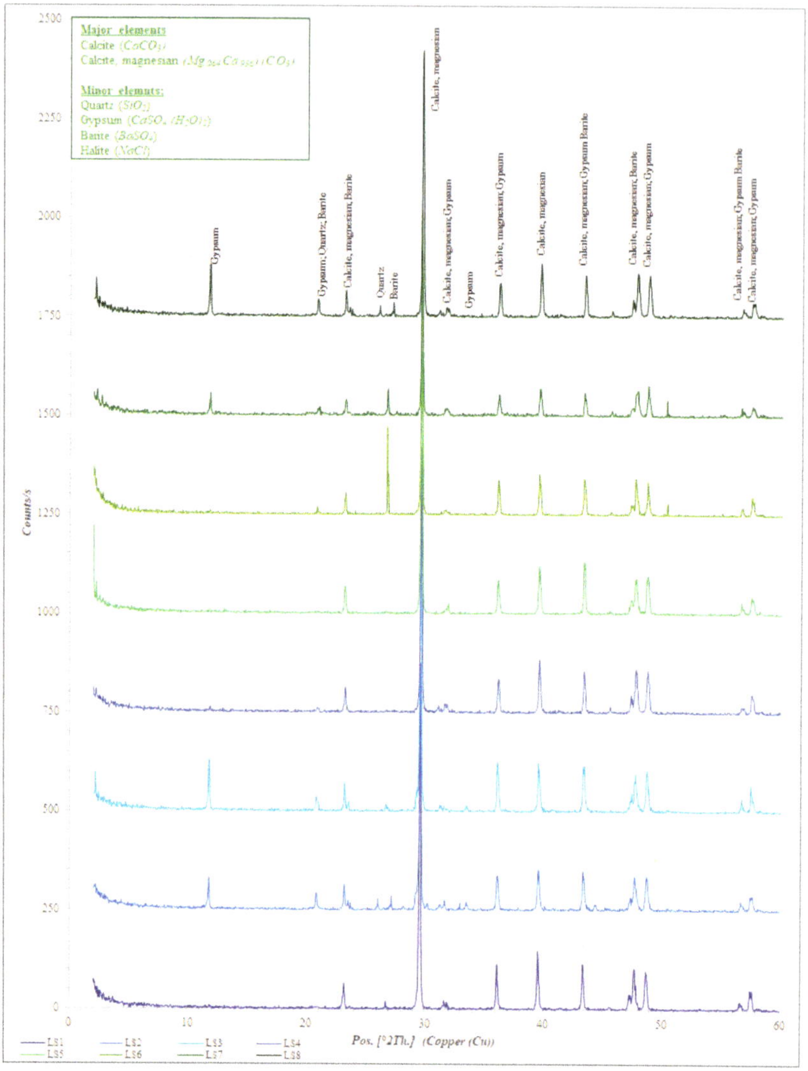

Figure 9. X-ray diffraction pattern of representative limestone sample.

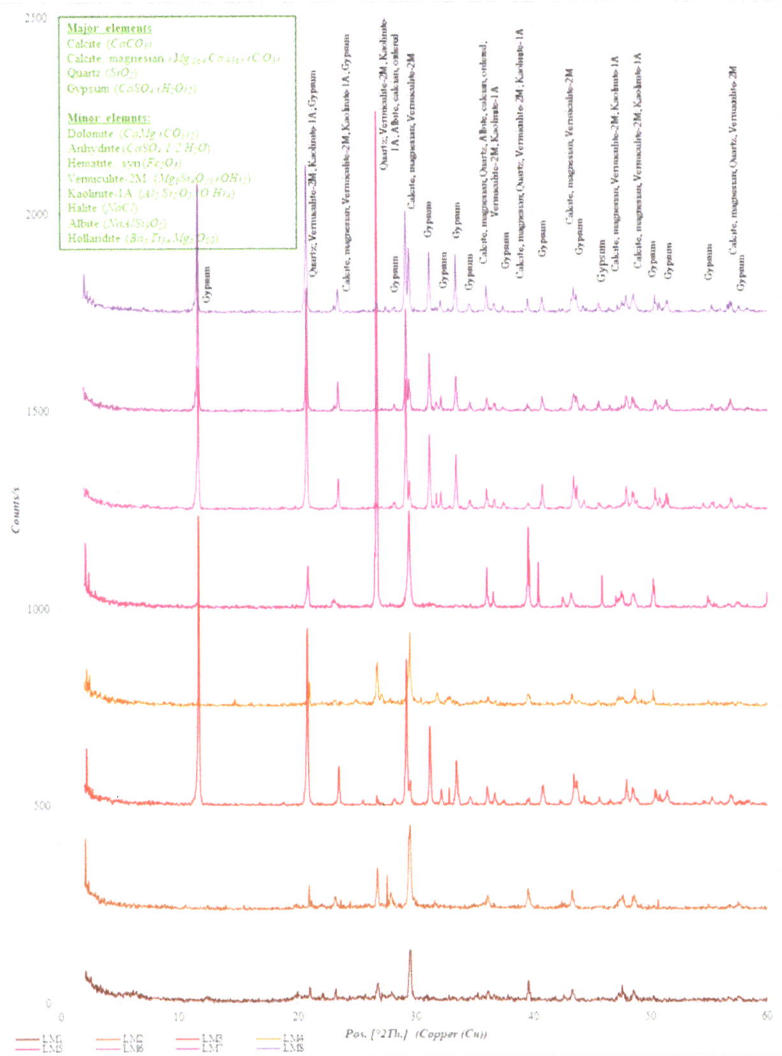

Figure 10. X-ray diffraction pattern of representative mortar sample.

3.4. Thermal Analysis (TG/DTA)

The results obtained by means of XRD have been confirmed by the thermal analyses carried out on the same investigated stone and mortar samples. According to the TGA/DTA analysis results related to stone samples, an initial weight loss of around 0.2–0.4% up to 110–120 °C is due to the release of absorbed hygroscopic water. Moreover, between 680 °C and 750 °C the obtained curve is partly erratic with little steps around 1.7–5.7%, this could be due to the presence of very small amounts of dolomite ($CaMg\ (CO_3)_2$) as dolomite decomposes in this temperature range [52,53]. The subsequent steps include weight loss of about 26–29.5% with decomposition at a temperature range of 650–860 °C, which can be attributed to the decarbonization of calcium carbonate (*Calcite*) $CaCO_3$. At 1000 °C, meanwhile, a total weight loss of around 20.7–28.6% was registered. The thermogravimetric analysis curves for studied stone samples are shown in Figure 11a, while Table 3 represents the mass losses (in %) obtained by TGA corresponding to the dehydration of hygroscopic water and the carbonates' decarbonization regions.

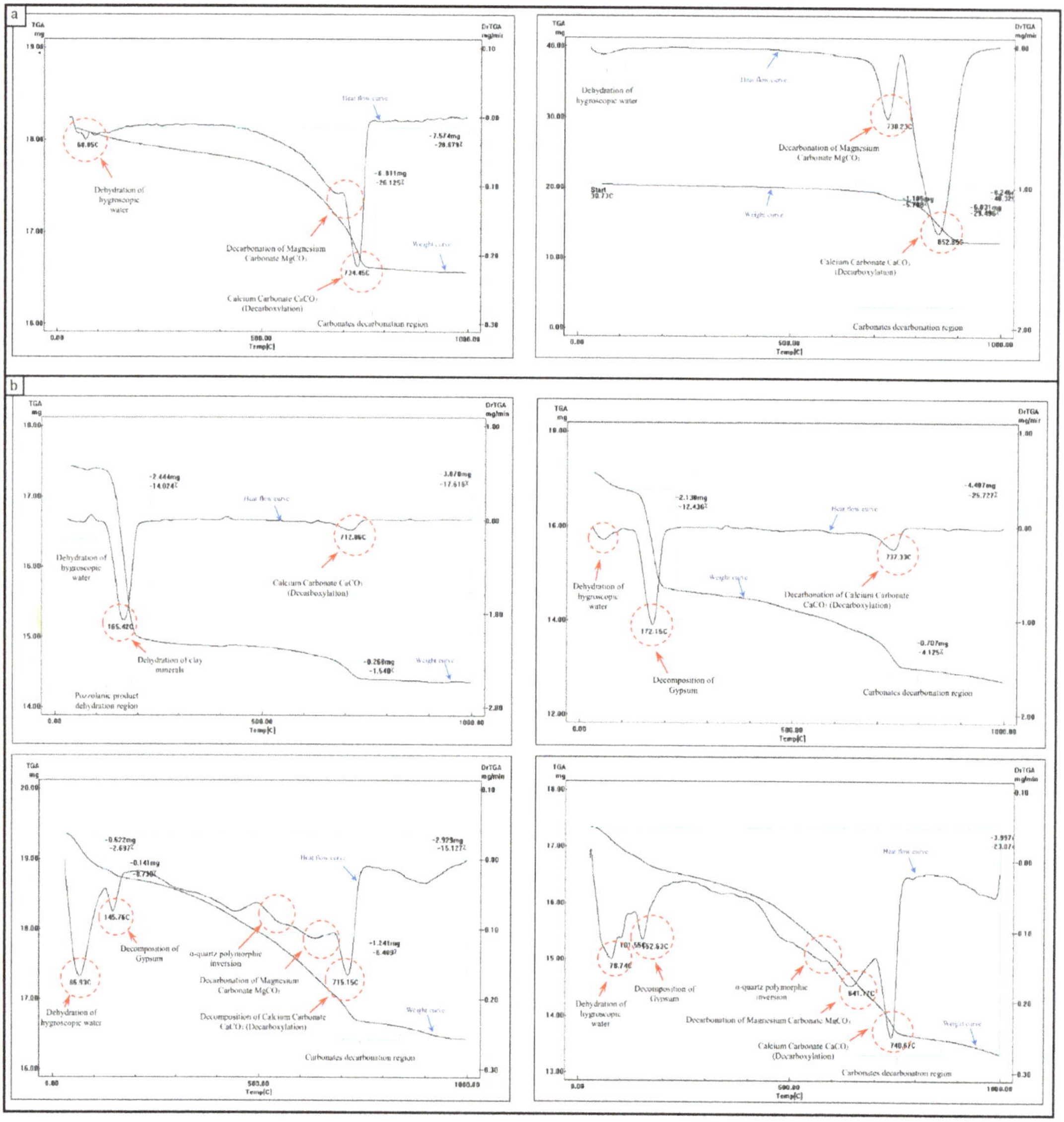

Figure 11. Thermogravimetric analysis curves of analysed (**a**) stone and (**b**) mortar sample.

Table 3. Results of DTA-TGA analysis of weight loss (in %) up to 1000 °C for stone and mortar samples.

Sample	Mass Loss %					Total Mass Loss %
	Dehydration of Hygroscopic Water	Decomposition of Gypsum	Dehydration of Clay Minerals	α-Quartz Polymorphic Inversion	Decomposition of Carbonates	
Stone	0.227	-	-	-	26.125	28.679
	0.432	-	-	-	35.294	40.329
Mortar	2.697	0.730	-	1.251	6.409	15.127
	2.135	12.436	-	-	4.125	25.727
	0.247	-	14.024	-	1.540	17.616
	2.542	1.834	-	1.254	7.351	23.074

Considering the investigated mortar samples are either from the external or internal layers, an initial weight loss of around 0.24–2.6% up to 120 °C is due to the release of absorbed hygroscopic water. In addition, endothermic peaks, with associated weight loss of about 0.7–12.4% in the range of 140 °C to 200 °C, have been registered. This peak could be ascribed to the loss of water of crystallization of the gypsum ($CaSO_4.2H_2O$). Gypsum dehydration takes place in two stages as follows (Equations (2) and (3)) [54,55]:

$$CaSO_4{\cdot}2H_2O \rightarrow CaSO_4{\cdot}\tfrac{1}{2}H_2O + \tfrac{3}{4}H_2O \quad \sim 145\ ^\circ C \tag{2}$$

$$CaSO_4{\cdot}\tfrac{1}{2}H_2O \rightarrow CaSO_4 + \tfrac{1}{2}H_2O \quad \sim 570\ ^\circ C \tag{3}$$

In some samples, the curve shows an initial endothermic peak at 100–250 °C, due to the dehydration of clay minerals as apparent representatives of the loss of water held between the basal planes of the lattice structure (i.e., swelling water) [56].

Furthermore, an endothermic peak almost without associated weight loss is observed. This peak at ~540–570 °C could be related to the polymorphic transformation of the α-Quartz. The peak is small because the energy associated with this change is minimal (Equation (4)) [53,57].

$$\alpha - quartz \rightarrow \beta - quartz \quad \sim 540\ ^\circ C \tag{4}$$

Besides, a point of inflection at ~620 °C to 680 °C accompanied by a weight loss of about 1.2–2.7%, can be observed and ascribed to dolomite decomposition. This association is in agreement with the XRD analysis results, while the weight loss of about 4.1–6.4% in the range 620–800 °C is attributed to the decarbonization of calcite $CaCO_3$ and the associated weight loss originates from CO_2 (g) evolution (Equation (5)) [52,53,58].

$$CaCO_3 \rightarrow CaO + CO_2 \quad \sim 715\ ^\circ C \tag{5}$$

Finally, at 1000 °C a total weight loss of around 15.1–25.5% was registered. The thermogravimetric analysis curves of analysed mortar samples are shown in Figure 11b.

3.5. Microstructure and Micro-Morphological Examination Using SEM with EDX

These SEM-EDX microanalyses are remarkably compatible with the previous microanalyses, where *Ca*, *Mg*, *C*, and *O* are components of carbonate phases $CaCO_3$, $Ca\ Mg\ (CO_3)_2$ (calcite and calcite magnesian) in XRD and thermal analyses; *Si* and *O* are components of quartz; *Fe* is a component of Fe_2O_3 (Hematite); *Cl* is a component of NaCl (Halite); *C*, *S*, and *O* are components of $CaSO_4.2H_2O$ and $CaSO_4.1/2H_2O$ (gypsum and anhydrite), respectively. While *Al*, *Si*, *P*, *K*, and *Fe*, are components of $Mg_3Si_4O_{10}\ (OH)_2$, $Al_2\ Si_2\ O_5\ (OH)_4$ (vermiculite and kaolinite), respectively. The presence of iron in the form of iron hydroxides in the fine limestone (micrite) determines the slightly reddish colour of the analysed limestone.

The microstructure and EDX microanalysis of various stone samples listed in Figure 12a showed that the dominant constituent of the stone samples is calcite with fine rounded quartz crystals and, in some cases, traces of gypsum. The presence of *Ca, O,* and *Mg* indicates the carbonate phases $CaCO_3$, $Ca\ Mg\ (CO_3)_2$, while *Ca, S,* and *O* are related to the infection of the stone sample by sulphate salt decay (gypsum) $CaSO_4$, $CaSO_4{\cdot}2(H_2O)$. Moreover, the presence of iron in the form of iron hydroxides in the fine limestone (micrite) determines the slightly reddish colour of the analysed limestone samples. In some samples, heterogeneous distribution of gaps and a minor quantity of crystalline salts (halite, *NaCl*) were noticed through the stone. Furthermore, EDX microanalysis of various mortar samples listed in Figure 12b showed that all of the analysed mortars are lime-based mortar, as the principal constituent of the mortar samples is calcite with fine rounded quartz crystals, and in most cases gypsum is detached as a minor element or even trace element; conversely, in a few cases, gypsum is detached as a significant element with calcite. Different percentages and sizes of quartz grains are scattered in the mortar.

Moreover, heavy decomposition and migration of calcite crystals were observed; besides, an inhomogeneous distribution of the gaps through the mortar was noticed. In some samples, these pores and cavities of the mortar surface were partially filled by sodium chloride precipitations as salt decay. The presence of *Ca, Mg, C,* and *O* indicates the Carbonate phases $CaCO_3$, $Ca\ Mg\ (CO_3)_2$, while *Na, S,* and *Cl* correspond to the infection by decayed chloride and sulphate salts (e.g., *NaCl* and $CaSo_4$).

Additionally, *Si* and *O* are related the quartz, while the high concentration of phosphate (P) is attributed to phosphates salts; other trace elements indicate the impurities in the mortar mixture. It is worth remarking that organic additives (i.e., chopped straw) were observed in some mortar samples derived from the inner-core layers; these organic additives were often used for strengthening the mortar and increasing its cohesion with the rubble stone to fill the inner core of the wall.

3.6. Physical Characterization

The results obtained from the experimental tests demonstrate that the limestone specimens exhibit the lowest porosity values with an average of 17.22%, while the lime-based mortar specimens, which represent the embedded joints, exhibit higher porosity values with an average of 24.90%. Moreover, the core-infill specimens corresponding to the inner layer of multiple-leaf masonry walls exhibit the highest values of porosity with an average of 31%. Furthermore, the limestone specimens exhibit the highest dry and bulk density values, while the lime-based mortar specimens exhibit lower values, and the core-infill specimens exhibit the lowest values. The average dry and bulk density values of limestone, mortar and core-infill specimens are 2.086, 1.811, 1.715 g/cm^3 and 2.258, 2.060, 2.025 g/cm^3, respectively. It is observed that the porosity values present some scattering opposite to density, whose average values are associated with a remarkably low coefficient of variation. The average values of the porosity (η), dry density (ρ_d), bulk density (ρ_{Bulk}), water absorption (WA), specific weight (G_s) and void ratio (e) obtained with reference to cubic specimens as well as the corresponding coefficients of variation (CV) are shown in Table 4.

Table 4. Average results for the physical tests of limestone, mortar and core-infill specimens.

	ρ_d		ρ_{Bulk}		WA		η		G_s		e	
	Avg. (g/cm^3)	*CV* %	*Avg.* (g/cm^3)	*CV* %	*Avg.* %	*CV* %	*Avg.* %	*CV* %	*Avg.* (g/cm^3)	*CV* %	*Avg.* %	*CV* %
Limestone	2.086	8.95	2.258	8.90	8.27	13.85	17.22	15.52	2.525	10.85	20.91	18.47
Lime mortar	1.811	16.00	2.060	14.71	13.97	38.05	24.90	38.87	2.444	18.73	34.99	48.98
Core-infill	1.715	2.72	2.025	4.74	18.10	30.76	31.00	30.23	2.528	14.94	47.45	47.34

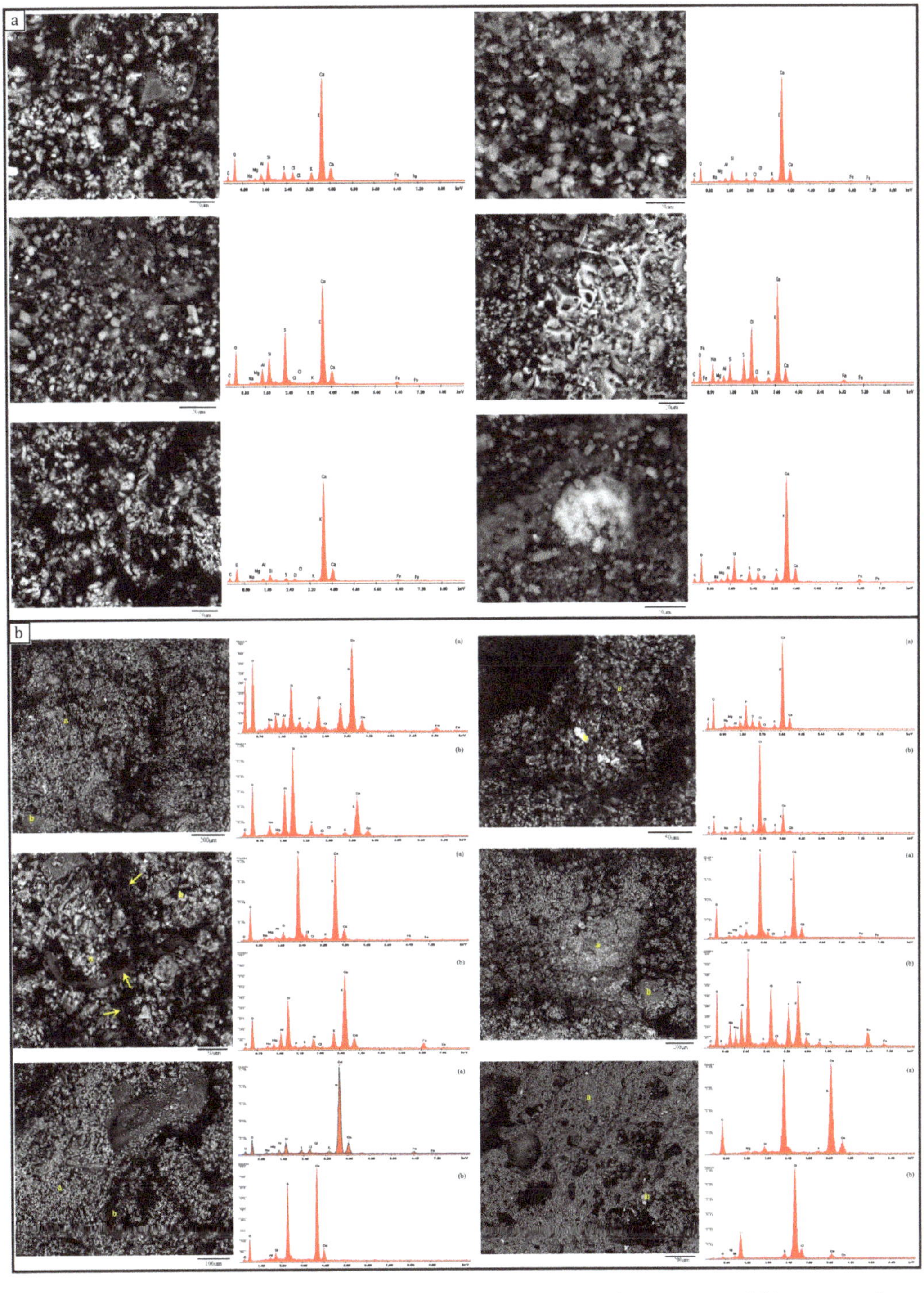

Figure 12. SEM photomicrographs and (EDX) microanalysis of thin sections of (**a**) limestone and (**b**) mortar samples.

3.7. Mechanical Characterization

In this section the results of the mechanical characterization of the materials are presented. Given the complexity of the topic, further details will be provided in a forthcoming paper. The notation adopted is in accordance with [7,27].

3.7.1. Uniaxial Compression Test

The average values of the compressive strength, f_c, peak strain, ε_p, initial cracking stress, $f_{c,i}$, initial cracking strain, ε_{ci}, initial elastic modulus, E_i, tangent elastic modulus, E_t, secant elastic modulus, E_s, obtained in the reference specimens tested under uniaxial compression are summarized in Table 5 and Figure 13. Examples of failure patterns of the tested specimens illustrating strain localization are shown in Figure 14.

Table 5. Average results obtained from the compression tests on cubic specimens of 150 × 150 × 150 mm, (values in brackets give the CV).

Tested Specimen	f_c N/mm^2	$f_{c,i}$ N/mm^2	ε_{ci} %	ε_p %	E N/mm^2	E_t N/mm^2	E_s N/mm^2
Limestone	22.72 (3.10)	6.44	0.11	0.41	6765.1 (7.05)	6666.67	6021.50
Lime-based mortar	1.66 (4.02)	1.23	-	-	1245	-	-
Core-infill	4.19 (11.02)	0.85	0.05	0.24	2350.47 (16.43)	1250	1142.85

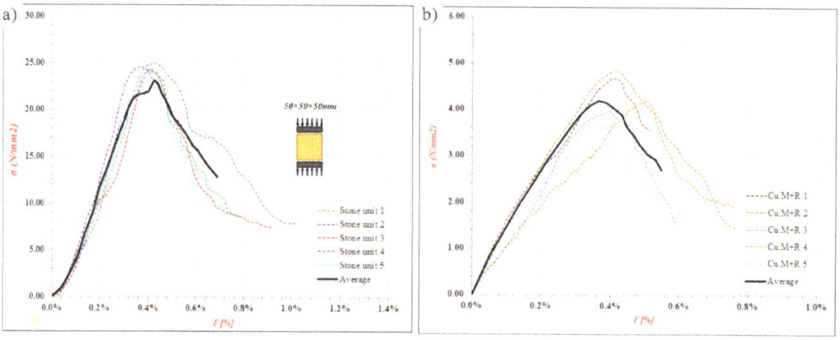

Figure 13. Stress–strain diagram obtained for limestone specimens with (**a**) limestone and (**b**) core-infill specimens.

Figure 14. Examples of failure patterns for tested (**A**) limestone, (**B**) lime-based mortar and (**C**) core-infill specimens under uniaxial compression.

Considering limestone units, the values found for the modulus of elasticity lie on a relatively low interval, and a mean value of 6.8 E + 3 N/mm^2 is attained. The compressive strength, $f_{c,b}$, shows the same tendency and exhibits a relatively low range of variation (17.59–24.93 N/mm^2).

Regarding the lime-based mortar, the average compressive strength of mortar specimens, $f_{m,c}$, was 1.66 N/mm^2 after 120 days. According to [59], the modulus of elasticity, E_m, could be calculated from (Equation (6)):

$$E_m = K_E f_{m,c} \quad (6)$$

where the recommended value for K_E is 1000, however, in the literature [60,61], the K_E value may range between 500 and 1000. For this reason, the value for K_E was considered equal to 750; consequently, the considered average value for the modulus of elasticity, E_m, is 1245 N/mm^2.

The values that were obtained from testing the core-infill specimens for the modulus of elasticity lie on a relatively high interval, the minimum (1176.89 N/mm^2) and the maximum (3416.06 N/mm^2). By averaging the results concerning all specimens, a mean value of 2.35×10^3 N/mm^2 is attained. The compressive strength, $f_{c,i}$, shows the same tendency and exhibits a relatively high range of variation (2.77–4.85 N/mm^2).

3.7.2. Splitting Tension Test

According to the obtained results, average splitting tensile strength, $f_{t,s}$, of 2.29, 0.148 and 0.374 N/mm^2 was determined for limestone, lime-based mortar and core-infill specimens, respectively. Concerning the ratio between the tensile and compressive strengths, a value of about 10% was found for limestone, while a value of 9% was found for lime-based mortar and the core-infill. Regarding the core-infill specimens, from the visual assessment, the observed failures in the indirect tensile tests were mainly due to the loss of adhesion between the lime mortar and the stone rubble (Figure 15).

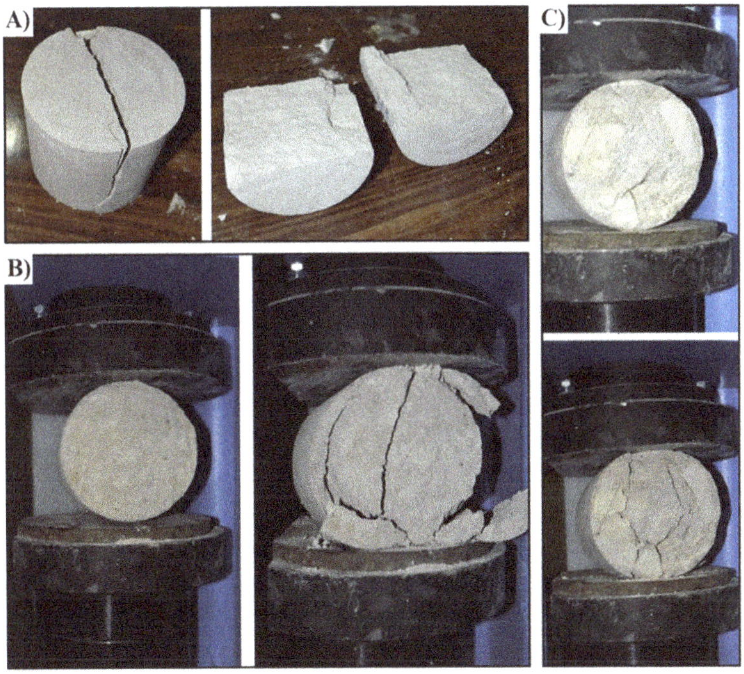

Figure 15. Failure mode of (**A**) limestone, (**B**) lime-based mortar, (**C**) core-infill specimens.

3.8. Thermal Conductivity and Resistivity

The results obtained from the conducted experimental tests demonstrate that the average thermal conductivity of limestone (corresponding to the outer layer) is about

0.365 W/mK, lime-based mortar (corresponding to the embedded joints) is about 0.549 W/mK, and core infill (corresponding to the inner layer) is about 0.402 W/mK. The density of the core infill increases when stone fragments are added so that the thermal resistivity increases by about 15.57–27.51%. Table 6 summarizes the results of thermal conductivity analysis of stone, mortar and core-infill.

Table 6. Thermal measurement results for the investigated limestone, mortar and core-infill specimens.

Specimens	Density, ρ (kg/m^3)	Temp. Upper °C	λ Upper (W/m K)	Temp. Lower °C	λ Lower (W/m K)	Percent Difference %	Mean Temp. °C	Avg. Thermal Conductivity, λ_{avg} (W/m K)
S1	2.2944	20.02	0.3244	45.03	0.3342	2.98	32.525	0.3293
S2	2.2763	20.02	0.3572	45.02	0.3319	7.34	32.52	0.3445
M1	1.6491	20.01	0.5671	45.02	0.5233	8.02	32.515	0.5452
M2	1.6016	20.01	0.5752	45.02	0.5308	8.02	32.515	0.5535
RM1	1.8525	20.02	0.3845	45.02	0.4178	8.29	32.520	0.4012
RM2	1.8477	20.01	0.4788	45.02	0.4418	8.02	32.515	0.4603

According to the thermal conductivity test results, it can be concluded that the thermal behaviour of stone masonry walls depends on various factors, mainly the density of their components and the void ratio. The tests results mentioned above proved that the thermal conductivity increases as the density increases. The limestone specimens with the highest density have the highest thermal resistivity, while the core-infill specimens have a lower density and thus present lower thermal resistivity. Additionally, lime-based mortar specimens, which have the lowest density, have the most significant thermal conductivity.

4. Conclusions

Concerning the experimental investigation on the construction materials used in constructing multiple-leaf stone-masonry walls, comprehensive microanalyses and testing programs were conducted to fully characterize their mineralogical, chemical and thermal properties. The results obtained from conducted microanalyses confirmed that:

- The outer leaves of the majority of the surveyed multiple-leaf stone-masonry walls in Egypt were mainly built of limestone blocks.
- The field survey results confirmed that most of the complex historical medieval buildings in Egypt present bearing structural elements built up by adopting multiple-leaf masonry technology. This building technology was used for vertical structural elements in almost all types of historical constructions, i.e., religious, service, residential, fortification, irrigation, etc. Moreover, multiple-leaf stone-masonry walls are characterized by different construction methods and typology that gradually changed from multiple-leaf walls with weak mechanical resistance made with a cohesionless internal core held by two separate external leaves to walls with fixed rubble-core masonry strongly connected and characterized by monolithic behaviour.
- The inner-core layer was built of rubble with bending mortar; this rubble is commonly consisting of rough and undressed limestone.
- Lime-based mortar is the most common type of mortar used in constructing both inner-core layer and external layers as a major binder between stone blocks in the case of external layers or between rubble-stones in the case of the inner-core layer.
- Mortar samples, collected either from the inner or external layers, are mainly composed of lime as the major binders, with sand as an aggregate and some additives used to enhance the adhesion performance of the mortar, such as red-brick powder (i.e., *Hommra*) or fly ash (i.e., *Qusrmil*) as pozzolanic materials. Sometimes gypsum $CaSO_4\,(H_2O)_2$ is found with varying percentages.
- According to the results obtained by means of thermogravimetric and differential thermal analyses, the temperature corresponding to the maximum decomposition rate

of the ancient lime mortar was 750 °C. Furthermore, the temperature corresponding to the maximum rate of decomposition of the historic limestone samples was 850 °C.

- According to the TGA/DTA analysis results of the lime-based mortar samples collected from the inner-core layer of different multiple-leaf stone-masonry walls, these mortars almost did not show any weight loss between 200 ° and 600 °C (related with the water of hydraulic compounds). Therefore, this could indicate that mortars have aerial lime as a binder. Moreover, some mortar samples have shown relatively low values of MgO. According to this fact, the use of magnesian-lime mortar could be proved. Moreover, the presence of magnesian calcite in the same samples has also been established by XRD.
- EDX microanalysis of various stone samples showed that the dominant constituent of the stone samples is Calcite with fine rounded Quartz crystals and, in some cases, traces of Gypsum. Additionally, the major constituent of the mortar samples is calcite with fine rounded quartz crystals. In most cases, gypsum is detected as a minor element or even trace element; conversely, in a very few cases, gypsum is detected as a significant element with calcite, particularly in walls of thicker cross-sections.
- In most cases, gypsum is detached as a minor element or even trace element in analysed mortar samples; conversely, gypsum is detached as a significant element with calcite in a few cases.
- The physical tests proved that the lime-based mortar of the embedded joints and limestone units of the outer leaves exhibit lower porosity values with an average of *24.9* and *17.2%*, respectively. In contrast, the inner core layer exhibits the highest porosity values with an average of *31%;* this is mainly due to the interfacial transition zone (ITZ) that exists between large particles of rubble stones and the hydrated lime-based mortar paste. Furthermore, the limestone specimens exhibit the highest dry and bulk density values, while the lime-based mortar specimens exhibit lower values, and the core-infill specimens exhibit the lowest values.
- Under uniaxial compression, a mean value of 6.8×10^3 N/mm^2 was attained for the modulus of elasticity of tested limestone specimens, while the mean values for the compressive strength obtained for limestone specimens, lime-based mortar, and core-infill cubic specimens after 120 days were 21.6, 1.6, 3.2 N/mm^2, respectively. The average values of the splitting tensile strength for limestone specimens, lime-based mortar, and core-infill specimens ranged from 9 to 11.6% of the corresponding compressive strength.
- The obtained failure pattern of core-infill specimens under compression and tension confirmed that the failure mode corresponds to the loss of adhesion between the lime mortar and the stone rubbles, i.e., a weak interfacial transition zone.
- According to the thermal conductivity test results, it can be concluded that the thermal behaviour of stone-masonry walls depends on various factors. The tests results proved that the thermal conductivity of multiple-leaf masonry walls depends mainly on the density of their components and the void ratio. Consequently, the thermal resistivity of the wall can be improved by decreasing the void ratio in the infill layer, increasing the cohesion between the bedding mortar and rubble stone, and also by using stones with lower permeability. Moreover, it is possible to infer that the increase in block thickness of the external layers and, above all, the use of mortar coating, attenuate the heat transfer to the inner layer of the wall.

Author Contributions: Conceptualization, O.A., D.A., E.K.M., A.T. and A.S.; methodology, O.A., D.A. and E.K.M.; investigation, O.A., A.T. and A.S.; visualization, O.A. and A.S.; writing—original draft preparation, O.A., E.K.M. and A.T.; writing—review and editing O.A., D.A. and A.S. All authors have read and agreed to the published version of the manuscript.

Funding: This research received no external funding.

Institutional Review Board Statement: Not applicable.

Informed Consent Statement: Not applicable.

Data Availability Statement: The data presented in this study are available on request from the corresponding author.

Acknowledgments: The authors heartily acknowledge Cairo University and the Housing & Building National Research Centre for their support. The authors are also grateful to Osama Talat, Amr Saleh, and Mustafa Bekhet for their collaboration and support.

Conflicts of Interest: The authors declare no conflict of interest.

References

1. Binda, L.; Penazzi, D.; Saisi, A. Historic masonry buildings: Necessity of a classification of structures and masonries for the adequate choice of analytical models. In Proceedings of the 6th International Symposium on Computer Methods in Structural Masonry (STRUMAS VI), Roma, Italy, 22–24 September 2003; pp. 168–173.
2. Binda, L.; Cardani, G.; Saisi, A. A classification of structures and masonries for the adequate choice of repair. In Proceedings of the International RILEM Workshop on Repair Mortars for Historic Masonry, Delft, The Netherlands, 26–28 January 2005; pp. 20–34.
3. Amer, O.; Abdel-Aty, Y.; Abdel-Hady, M.; Aita, D.; Torky, A.; Hussein, Y. Multiscientific-based approach to diagnosis and characterization of historic stone-masonry walls: The mausoleum of al-imam al-shafi'i, cairo (egypt). *Mediterr. Archaeol. Archaeom.* **2020**, *20*, 1–16.
4. Giamundo, V. Seismic Assessment and Retrofit of Historical Masonry Barrel Vaults. Ph.D. Thesis, University of Naples Federico II, Naples, Italy, 2014.
5. Bakeer, T. Collapse Analysis of Masonry Structures under Earthquake Actions. Ph.D. Thesis, Technischen Universität Dresden, Dresden, Germany, 2008.
6. Drysdale, R.G.; Hamid, A.A.; Baker, L.R. *Masonry Structures: Behavior and Design*; de leon, B.M., Zurite, P., Handy, S., Eds.; Masonry Society, Prentice Hall: Englewood Cliffs, NJ, USA, 1999.
7. Silva, B.Q.; Pappas, A.; Guedes, J.M.; da Porto, F.; Modena, C. Numerical analysis of the in-plane behaviour of three-leaf stone masonry panels consolidated with grout injection. *Bull. Earthq. Eng.* **2017**, *15*, 357–383. [CrossRef]
8. Anzani, A.; Binda, L.; Fontana, A.; Henriques, J.P. An experimental investigation on multiple-leaf stone masonry. In Proceedings of the 13th International Brick and Block Masonry Conference, Amsterdam, The Netherlands, 4–7 July 2004.
9. Pina-Henriques, J.; Lourenço, P.; Binda, L.; Anzani, A. Testing and modelling of multiple-leaf masonry walls under shear and compression. In *Structural Analysis of Historical Constructions*; Modena, C., Lourenço, P.B., Roca, P., Eds.; Taylor & Francis Group: London, UK, 2004; pp. 299–310. ISBN 04 1536 379 9.
10. Capozucca, R. Historic multiple-leaf masonry wall models under compression and cyclic shear loads. In *Structural Analysis of Historic Construction*; D'Ayala, D., Fodde, E., Eds.; Taylor & Francis Group: London, UK, 2008; pp. 297–302.
11. Magenes, G.; Penna, A.; Galasco, A.; da Paré, M. In-plane cyclic shear tests of undressed double-leaf stone masonry panels. In Proceedings of the 8th International Masonry Conference, Dresden, Germany, 4–7 July 2010.
12. Silva, B.; Pigouni, A.E.; Valluzzi, M.R.; Modena, C. Calibration of analytical formulations predicting compressive strength in consolidated three-leaf masonry walls. *Constr. Build. Mater.* **2014**, *64*, 28–38. [CrossRef]
13. Elmenshawi, A.; Shrive, N. Assessment of multi-wythe stone masonry subjected to seismic hazards. *J. Earthq. Eng.* **2015**, *19*, 85–106. [CrossRef]
14. Aldreghetti, I.; Baraldi, D.; Boscato, G.; Cecchi, A.; Reccia, E.; Massaria, L.; Tofani, I. Damage-imperfection indicators for the assessment of multi-leaf masonry walls under different conditions. In Proceedings of the 10th International Masonry Conference, Milan, Italy, 9–11 July 2018.
15. Vintzileou, E.; Miltiadou-Fezans, A.; Vrouva, A.; Anagnostopoulou, S. Mechanical Properties of Three-Leaf Stone Masonry. In *Structural Analysis of Historical Constructions*; Lourenço, P., Roca, P., Modena, C., Agrawal, S., Eds.; Springer: New Delhi, India, 2006; pp. 783–790.
16. Lombillo, I.; Thomas, C.; Villegas, L.; Fernández-Álvarez, J.; Norambuena-Contreras, J. Mechanical characterization of rubble stone masonry walls using non and minor destructive tests. *Constr. Build. Mater.* **2013**, *43*, 266–277. [CrossRef]
17. Demir, C.; Ilki, A. Characterization of the materials used in the multi-leaf masonry walls of monumental structures in Istanbul, Turkey. *Constr. Build. Mater.* **2014**, *64*, 398–413. [CrossRef]
18. Meimaroglou, N.; Mouzakis, H. Mechanical properties of three-leaf masonry walls constructed with natural. *Eng. Struct.* **2018**, *172*, 869–876. [CrossRef]
19. Valluzzi, M.R.; Mazzon, N.; Munari, M.; Casarin, F.; Modena, C. Effectiveness of injections evaluated by sonic tests on reduced scale multi-leaf masonry building subjected to seismic actions. In Proceedings of the NDTCE'09, Non-Destructive Testing in Civil Engineering, Nantes, France, 30 June–3 July 2009.
20. Oliveira, D.V.; Silva, R.A.; Garbin, E.; Lourenco, P.B. Strengthening of three-leaf stone masonry walls: An experimental research. *Mater. Struct.* **2012**, *45*, 1259–1276. [CrossRef]
21. Gemert, D.V.; Ignoul, S.; Brosens, K.; Toumbakari, E.-E. Consolidation and strengthening of historical masonry by means of mineral grouts: Grout development. *Restor. Build. Monum.* **2015**, *21*, 29–45. [CrossRef]

22. Capozucca, R. Double-leaf masonry walls under in-plane loading strengthened with GFRP/SRG strips. *Eng. Struct.* **2016**, *128*, 453–473. [CrossRef]
23. Candela, M.; Borri, A.; Corradi, M.; Righetti, L. Effect of transversal steel connectors on the behaviour of rubble stonemasonry walls: Two case studies in Italy. In Proceedings of the 16th International Brick and Block Masonry Conference, Padova, Italy, 26–30 June 2016.
24. Nikolopoulou, V.; Adami, C.-E.; Karagiannaki, D.; Vintzileou, E.; Miltiadou-Fezans, A. Grouts for strengthening two- and three leaf stone masonry, made with earthen mortars. *Int. J. Archit. Herit.* **2018**, *13*, 663–678. [CrossRef]
25. Mahmoud, H.S. Multiscientific approach for the characterization and assessment of the degradation state of the historical Al-Shafi'i mosque walls (Jeddah, Kingdom of Saudi Arabia). *Sci. Cult.* **2021**, *7*, 1–19.
26. Salama, K.K.; Ali, M.F.; Moussa, A.M. The presence of cement mortars in the added chambers of El Sakakeny Palace: A case study. *Sci. Cult.* **2017**, *3*, 25–29.
27. Amer, O. Experimental and Analytical Studies on Structural Behavior of Multiple-Leaf Masonry Walls under Loads and Biological Factors, and the Appropriate Restoration Techniques with Application on Chosen Historical Islamic Buildings Biologically Affect. Master's Thesis, Conservation Department, Faculty of Archaeology, Cairo University, Cairo, Egypt, 2018.
28. Nuffield, E.W. *X-ray Diffraction Methods*; Wiley: Hoboken, NJ, USA, 1966.
29. Warren, B.E. *X-ray Diffraction*; Courier Corporation: Chelmsford, MA, USA, 1969.
30. Suryanarayana, C.; Norton, M.G. *X-ray Diffraction: A Practical Approach*; Springer Science, Business Media: New York, NY, USA, 1998.
31. Williams, C.; May, R.P.; Guinier, A. *Characterization of Materials*; Lifshin, E., Ed.; Wiley-VCH: Weinheim, Germany, 1999.
32. Waseda, Y.; Matsubara, E.; Shinoda, K. *X-ray Diffraction Crystallography: Introduction, Examples and Solved Problems*; Springer: Berlin/Heidelberg, Germany, 2011.
33. Patel, K.P. Crystal of Metal NI(II) and CU(II) in Salicyldehyde and Ethelenediamine Solution: Preparation and Characterization. Ph.D. Thesis, Department of Physics, Suresh Gyan Vihar University, Jaipur Rajasthan, India, 2011.
34. Panda, S.S.; Mohapatra, P.K.; Chaturvedi, R.K.; Kar, S.K. Chemical analysis of ancient mortar from excavation sites of Kondapur, Andhra Pradesh, India to understand the technology and ingredients. *Curr. Sci.* **2013**, *105*, 837–842.
35. Theologitis, A.; Kapridaki, C.; Kallithrakas-Kontos, N.; Maravelaki-Kalaitzaki, P.; Fotiou, A. Mortar and plaster analysis as a directive to the design of compatible restoration materials in frangokastello (crete). *Mediterr. Archaeol. Archaeom.* **2021**, *21*, 109–120.
36. al Sekhaneh, W.; Shiyyab, A.; Arinat, M.; Gharaibeh, N. Use of FTIR and thermogravimetric analysis of ancient mortar from The Church of the Cross in Gerasa (Jordan) for conservation purposes. *Mediterr. Archaeol. Archaeom.* **2020**, *20*, 159–174.
37. Hatakeyama, T.; Quinn, F. *Thermal Analysis: Fundementals and Applications to Polymer Science*; John Wiley & Sons Ltd.: Hoboken, NJ, USA, 1994.
38. Jones, C. *The Use of Engineering Technology in the Determination of Historic Brickwork*; Honors Summer Research, Tusculum Institute, Sweet Briar College: Sweet Briar, VA, USA, 2011.
39. ISRM. *Suggested Methods for Determining Water Content, Porosity, Density, Absorption and Related Properties and Swelling and Slake-Durability Index Properties*; Brown, E.T., Ed.; Pergamon: Oxford, UK, 1981.
40. ASTM D2938-95. *Standard Test Method for Unconfined Compressive Strength of Intact Rock Core Specimens*; ASTM International: West Conshohocken, PA, USA, 2002.
41. ISRM. *Suggested Method for Determining Uniaxial Compressive Strength and Deformability of Rock Materials*; Brown, E.T., Ed.; Pergamon: Oxford, UK, 1981.
42. Kržan, M. Performance Based Experimental and Numerical Assessment of Multi-Leaf Stone Masonry Walls. Ph.D. Thesis, Fakulteta za Gradbeništvo in Geodezijo, Univerza v Ljubljani, Ljubljana, Slovenia, 2015.
43. ASTM C518-10. *Standard Test Method for Steady-State Thermal Transmission Properties by Means of the Heat Flow Meter Apparatus*; ASTM: West Conshohocken, PA, USA, 2003.
44. EN ISO 12667. *Thermal Performance of Building Materials and Products-Determination of Thermal Resistance by Means of Guarded Hot Plate and Heat Flow Meter Methods-Products of High and Medium Thermal Resistance*; ISO: Geneva, Switzerland, 2001.
45. ISO 8301. *Thermal Insulation Determination of Steady-State Thermal Resistance and Related Properties Heat Flow Meter Apparatus*; ISO: Geneva, Switzerland, 1991.
46. Yüksel, N. The Review of Some Commonly Used Methods and Techniques to Measure the Thermal Conductivity of Insulation Materials. In *Insulation Materials in Context of Sustainability*; Almusaed, A., Almssad, A., Eds.; IntechOpen: London, UK, 2016; pp. 113–140. ISBN 978-953-51-2624-9.
47. Buratti, C.; Moretti, E.; Belloni, E.; Agosti, F. Development of Innovative Aerogel Based Plasters: Preliminary Thermal and Acoustic Performance Evaluation. *Sustainability* **2014**, *6*, 5839–5852. [CrossRef]
48. Giuffré, A. Mechanics of historical masonry and strengthening criteria. In Proceedings of the XV Regional Seminar on Earthquake Engineering, Ravello, Italy, 18–23 September 1989.
49. Pulatsu, B.; Bretas, E.M.; Lourenco, P.B. Discrete element modeling of masonry structures: Validation and application. *Geomech. Eng.* **2016**, *11*, 563–582. [CrossRef]
50. Pulatsu, B.; Gencer, F.; Erdogmus, E. Study of the effect of construction techniques on the seismic capacity of ancient dry-joint masonry towers through DEM. *Eur. J. Environ. Civ. Eng.* **2020**, 1–18. [CrossRef]

51. Frank-Kamenetskaya, O.V.; Vlasov, D.Y.; Zelenskaya, M.S.; Knauf, I.V.; Timasheva, M.A. Decaying of the marble and limestone monuments in the urban environment. Case studies from Saint Petersburg, Russia. *Studia Univ. Babeş-Bolyai Geol.* **2009**, *54*, 17–22. [CrossRef]
52. Webb, T.; Krüger, J. *Differential Thermal Analysis*; Mackenzie, R., Ed.; Academic Press Inc.: Cambridge, MA, USA, 1970; Volume 1, pp. 303–341.
53. Montoya, C.; Lanas, J.; Arandigoyen, M.; Casado, P.G.; Alvarez, J. Mineralogical, chemical and thermal characterisations of ancient mortars of the church of Santa María de Irache monastery (Navarra, Spain). *Mater. Struct.* **2004**, *37*, 433–439. [CrossRef]
54. Hatakeyama, T.; Liu, Z. *Handbook of Thermal Analysis*; John Wiley & Sons Ltd.: Hoboken, NJ, USA, 1998.
55. Singh, N.B.; Middendorf, B. Calcium sulphate hemihydrate hydration leading to gypsum crystallization. *Prog. Cryst. Growth Charact. Mater.* **2007**, *53*, 57–77. [CrossRef]
56. Gnru, R.E.; Rowlawp, R.A. Differential thermal analyses of clay minerals and other hydrous materials. Part 1. *Am. Mineral. J. Earth Planet. Mater.* **1942**, *27*, 746–761.
57. Newton, R.; Sharp, J. The chemical composition of lime plasters. *Cem. Concr. Res.* **1987**, *18*, 151–155. [CrossRef]
58. Alvarez, J.; Navarro, I.; Casado, P.G. Thermal, mineralogical and chemical studies of the mortars used in the cathedral of Pamplona (Spain). *Thermochim. Acta* **2000**, *365*, 177–187. [CrossRef]
59. CEN. *Eurocode 6: Design of Masonry Structures*; EN 1996-1-1:2003; CEN: Brussels, Belgium, 2003.
60. Vasconcelos, G. Experimental Investigations on the Mechanics of Stone Masonry: Characterization of Granites and Behavior of Ancient Masonry Shear Walls. Ph.D. Thesis, University of Minho, Braga, Portugal, 2005.
61. Lourenço, P.; Hees, R.; Lubelli, F.F.B. Characterization and Damage of Brick Masonry. In *Structural Rehabilitation of Old Buildings*; Costa, A., Guedes, J.M., Varum, H., Eds.; Springer: Berlin/Heidelberg, Germany, 2014; pp. 9–130.

heritage

Article

Multi-Leaf Stone Masonry Walls in Egypt: A Legend

Osama Amer [1,*], Danila Aita [2], Ezzeldin k. Mohamed [3], Akram Torky [4] and Ashraf Shawky [4]

1 Department of Conservation, Cairo University, Giza 12613, Egypt
2 Architecture and Design Department, University of Genoa, 16123 Genoa, Italy; danila.aita@edu.unige.it
3 National Water Research Centre, Construction Research Institute, Shoubra El-Kheima 13621, Egypt; ezzeldin_kamel@nwrc.org.eg
4 Department of Civil Engineering, Faculty of Engineering, Cairo University, Giza 12613, Egypt; amtorkey@eng.cu.edu.eg (A.T.); ashraf.shawky@mentor-mcg.com (A.S.)
* Correspondence: osama_amer@cu.edu.eg

Abstract: Multi-leaf stone masonry walls are a typical construction technique in architectural heritage in Egypt. The assessment, modeling, and strengthening of historic masonry walls of multi-leaf systems essentially require suitable knowledge of their construction technology, typology, geometrical characteristics, and the properties of their components. Within the current research project, a comprehensive structural survey of multiple-leaf walls of medieval historic buildings in Cairo was performed. The observation and statistical analyses allowed characterization of the transversal section of the surveyed walls, as well as examination and identification of the construction materials and techniques. The slenderness ratios of this type of wall, its blocks' dimensions, the utilized connectivity between the inner and outer leaves, and leaves ratio were also investigated. Three construction hypotheses of multiple-leaf stone masonry walls are presented considering weak, thick, and monolithic core infill layers. The study's objectives were to enlarge the knowledge of typology, morphology, and construction materials used in three-leaf masonry walls and provide a proper characterization as a prerequisite for determining the most suitable materials and techniques for further strengthening interventions.

Keywords: historic constructions; three-leaf masonry walls; field survey; architectural heritage; construction history

Citation: Amer, O.; Aita, D.; Mohamed, E.k.; Torky, A.; Shawky, A. Multi-Leaf Stone Masonry Walls in Egypt: A Legend. *Heritage* **2021**, *4*, 2763–2791. https://doi.org/10.3390/heritage4040156

Academic Editor: Giuliana Cardani

Received: 17 August 2021
Accepted: 24 September 2021
Published: 28 September 2021

1. Introduction

Multi-leaf masonry walls (MLMW) constitute the construction typology most widely adopted in historic buildings worldwide. They are found, for example, in Roman buildings [1]. Most complex historic constructions in Egypt present bearing structural elements built up adopting the multiple-leaf masonry technique, which gives a non-homogeneous structural element. This building technology is around 4000 years old. It exists in a variety of forms [2]: it usually was used as vertical structural elements in historic houses, public buildings, religious constructions, and hydraulic structures. In more recent years, this type of wall was also adopted for piers in road and railway bridges [3–5].

Multiple-leaf stone masonry walls are characterized by different construction techniques and typologies that have gradually changed from those with weak and cohesionless inner core layers to the fixed rubble-core masonry. Furthermore, the inner and outer layers could be strongly bonded together utilizing transversal bond elements; in some cases, the separated behavior of each layer is attained due to the absence of the shear keys/through stones between the leaves (walls with straight collar joints) [6–8]. Many structural problems are encountered in these structures, due to a weak internal layer, mortar deterioration, and lack of connection between the leaves [8–10] (see Figure 1). These problems can lead to the development of high stresses and the separation of the leaves and, consequently, affect the overall stability of the wall [11,12].

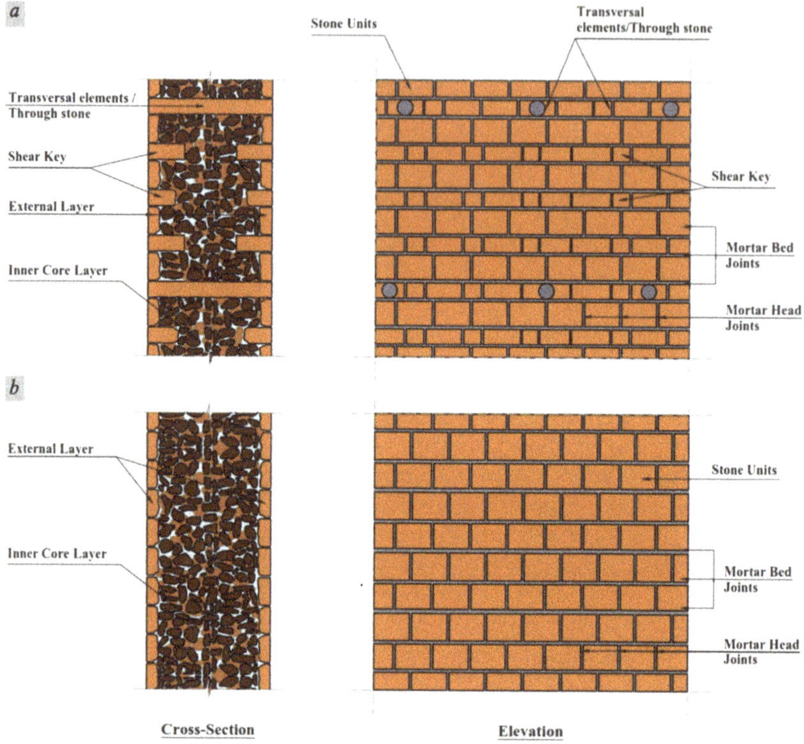

Figure 1. Common typologies of the of multiple-leaf stone masonry walls with (**a**) keyed and (**b**) straight collar joints.

Several monuments consisting of multi-leaf walls have experienced excessive damage and failure caused by high compressive and shear loads (e.g., *Bab Qaytbay*, *Takiyya Ibrahim al-Kulshani*, *Wakala al-Sultan Qaytbay*, among others). However, structural interventions can be properly accomplished in this type of construction if the properties and the existing internal defects or damage are properly identified. In addition, retrofitting/strengthening techniques—including grout injection where mortar is heavily deteriorated, mortar repointing of the external layers of the joints, and establishment of proper connection between the leaves—often alter the properties of the wall, increasing the stiffness of the inner layer and the wall itself and, thus, modifying its response to different loading conditions [13–18]. Undoubtedly, these interventions require thorough and accurate studies of the mechanical response of this type of construction before and after interventions to achieve the greatest possible compatibility.

Recently, some scholars have studied the mechanical behavior of the multi-leaf masonry walls, by examining material discretization and the role of each leaf in the global behavior of the wall [7,19–25]. Since the activities of old masons strongly depended on the available material near the construction site and on their experience, which was not handed down by written documents, it is not possible to find unified, simple design criteria relating the wall height, thickness, and filling material with reference to the function/type of the structure. Discovering this secret is a challenge. This was the motivation for investigating multi-leaf walls in historic constructions in Egypt during *Fatimid*, *Ayyubid*, *Mamluk*, and *Othman* eras and the development of this construction technique during the *Mohamed Ali* period and the British colony years.

In order to accurately assess the structural performance of this particular type of historical masonry wall under various types of loads, it is necessary to adequately understand the structure, its typology, its behavior and response, as well as the constituent materials and their mechanical characteristics. Moreover, the study of the multiple-leaf masonry wall sections can be significant with respect to structural analysis, by providing essential input data concerning wall morphology and geometry, which are useful for adopting appropriate mathematical models, formulating valid hypotheses on mechanical stability, and verifying possible failure mechanisms. Correspondingly, it is imperative to follow a methodology that includes a detailed survey and diagnosis to obtain reliable documentation corroborating the selection of the correct and effective intervention technique fitting best with the strengthening requirements. As a result, a detailed understanding of the morphology, typology, and utilized construction techniques of this common type of wall is necessary to minimize restoration interventions for strengthening or repair. This is the point of view from which the present paper developed some reflection about the construction aspects of the multiple-leaf stone masonry walls of the medieval constructions of architectural heritage in Egypt.

A systematic survey concerning the evolution of building materials and construction technology of multiple-leaf masonry load-bearing walls in Cairo during different medieval periods was carried out to analytically determine the construction techniques and structural characteristics, not clearly inferable from the literature. A total of thirty-three historic buildings in Cairo dating to different centuries of the Middle Ages and at various locations were surveyed. Various samples of stone blocks and mortars were studied for material characterization, while the research sources provided historical documentation on some surveyed buildings. The survey data are listed according to the historical period, including building name, type, period, current condition, usage, monument registration number (RN), and geometrical data of the multiple-leaf masonry walls. For further historical research on the surveyed building, the reader is referred to the following studies [26–42]. The monuments surveyed during the present study, summarized in Table 1 and Figure 2, included the following: two masjids, two main gates, four towers, five mausoleums, eleven complex buildings (i.e., *madrasa*, *khanqah*, *sabil-kuttab*, etc.), four palaces, and five other ruined monuments.

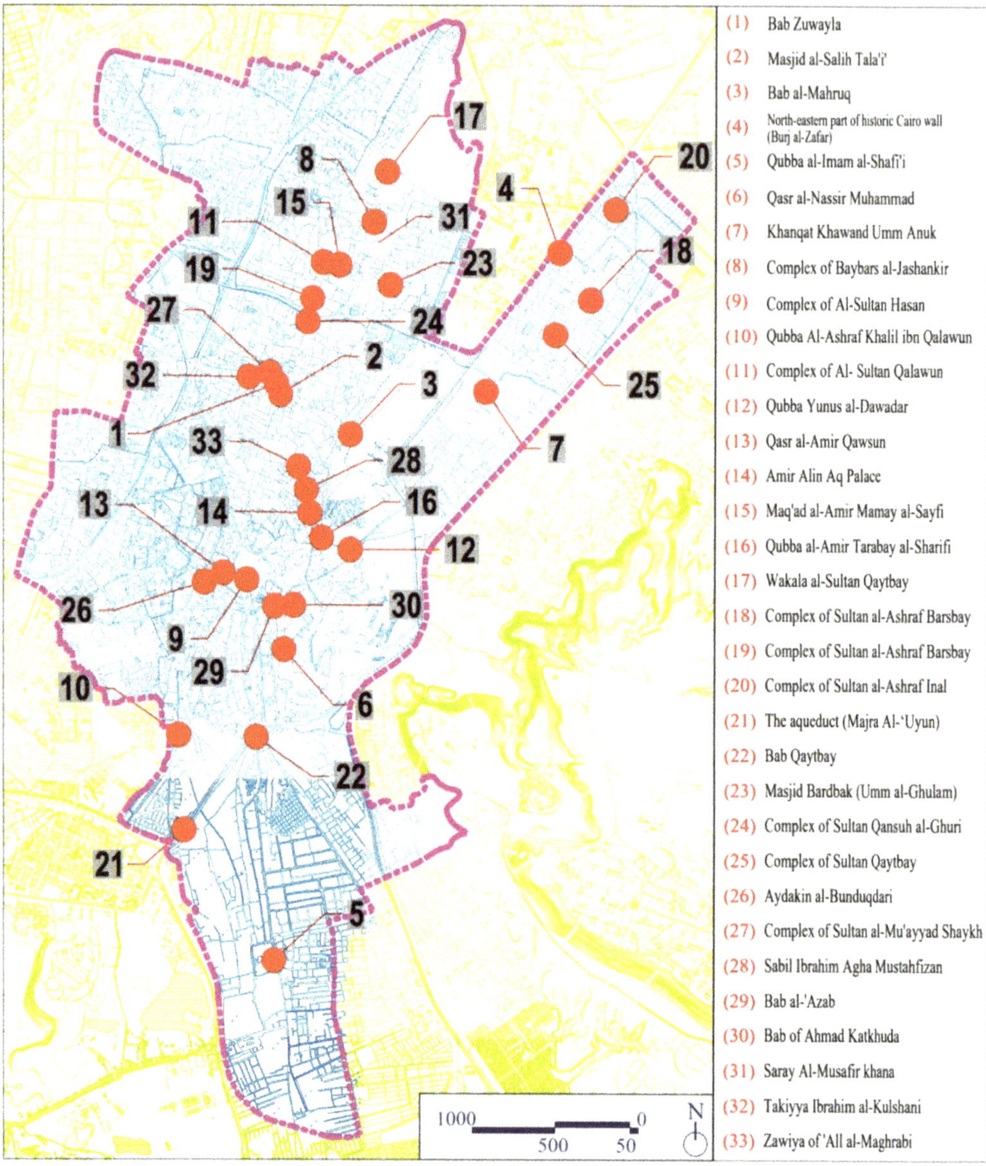

Figure 2. Locations of the surveyed buildings in historic Cairo.

Table 1. Summary of the conducted field surveyed.

Building	Monument Registration Number	Period	Type	Usage	Condition	Structure Material	Characteristics of Multiple Leaf Masonry Walls: Wall Thickness		Width of Leaves		Leaves Ratio	Slenderness Ration (λ)	Core Ratio % int./W.	
							From	to	ext.	int.	int./ext.			
Bab Zuwayla	199	Fatimid (AD 969–1171)	Military	Gate	V. good	Limestone, lime-based mortar, rubble infill	1.2	5	0.20–0.25	1–4.6	5–20	3.5–10.0	83.3	92.0
Masjid al-Salih Tala'i'	116		Religious	Masjid	Good	Limestone, lime-based mortar, rubble infill	1	1.4	0.20–0.25	0.8–1	4–5	7.8–11	80.0	71.4
Bab al-Mahruq	307	Ayyubid (AD 1171–1250)	Military	Tower	V. good	Limestone, lime-based mortar, rubble infill	0.8	2.5	0.20–0.25	0.4–2	2–8	4.7–8.0	50.0	80.0
North-eastern part of historic Cairo wall (Burj al-Zafar)	307		Military	Tower	Fair	Limestone, lime-based mortar, rubble infill	0.6	4.1	0.20–0.25	0.2–3.6	1.0–15	3.0–8.0	33.3	87.8
Qubba al-Imam al-Shafi'i	281		Funerary	Mausoleum	Fair	Limestone, lime-based mortar, rubble infill	2.6	2.8	0.23–0.25	1.1–2.3	8.4–10	3.57–3.94	42.3	82.1
Qasr al-Nassir Muhammad	549	Bahri Mamluk (AD 1250–1382)	Palatial	Palace	Ruins	Limestone, bricks, lime-based mortar, rubble infill	0.7	1.95	0.20–0.25	0.2–1.6	1.5–6.5	4.0–6.5	28.6	82.1
Khanqat Khawand Umm Anuk	81		Funerary	Mausoleum	Ruins	Limestone, bricks, lime-based mortar, rubble infill	1	3	0.20–0.25	0.6–2.5	3–10	3.0–4.5	60.0	83.3
Complex of Baybars al-Jashankir	32		Religious, Funerary	Mausoleum, Khanqah	Fair	Limestone, lime-based mortar, rubble infill	1	3.3	0.20–0.25	0.6–2.8	3–11.5	4.0–12.0	60.0	84.9
Complex of Al-Sultan Hasan	133		Religious, Educational, Funerary	Masjid, Madrasa, Mausoleum	V. good	Limestone, lime-based mortar, rubble infill	1.2	5	0.20–0.30	0.8–4.5	4–18	6.5–20.5	66.7	90.0

Table 1. *Cont.*

Building	Monument Registration Number	Period	Type	Usage	Condition	Structure Material	Characteristics of Multiple Leaf Masonry Walls: Wall Thickness From	Wall Thickness to	Width of Leaves ext.	Width of Leaves int.	Leaves Ratio int./ext.	Slenderness Ration (λ)	Core Ratio % int./W.	
Qubba Al-Ashraf Khalil ibn Qalawun	275		Funerary	Mausoleum	Good	Limestone, lime-based mortar, rubble infill	1.4	1.9	0.20–0.25	1–1.4	5–7	5.3–7.2	71.4	73.7
Complex of Al- Sultan Qalawun	43		Educational, Health-care, Religious	Masjid, Hospital, Madrasa	V. good	Limestone, lime-based mortar, rubble infill	1.2	4.8	0.20–0.30	0.8–4.3	4–21	4.0–10.5	66.7	89.6
Qubba Yunus al-Dawadar	139		Funerary	Mausoleum	Ruins	Limestone, lime-based mortar, rubble infill	0.7	1.2	0.20–0.25	0.3–0.8	1.5–4	–	42.9	66.7
Qasr al-Amir Qawsun	266		Palatial	Palace	Ruins	Limestone, bricks, lime-based mortar, rubble infill	1.1	4.4	0.20–0.35	0.4–2.2	2–8.5	4.0–11.0	36.4	50.0
Amir Alin Aq Palace	249		Palatial	Palace	Good	Limestone, bricks, lime-based mortar, rubble infill	1.3	3	0.20–0.25	0.9–2.5	4.5–10	4.0–9.5	69.2	83.3
Maq'ad al-Amir Mamay al-Sayfi	51	Circassian Mamluk (AD 1382–1517)	Residential	Shelter	Good	Limestone, lime-based mortar, rubble infill	0.6	1.8	0.20–0.25	0.2–1.4	1–7.0	10.0–30.0	33.3	77.8
Qubba al-Amir Tarabay al-Sharifi	255		Funerary	Mausoleum	V. good	Limestone, lime-based mortar, rubble infill	0.5	1.5	0.20–0.25	0.08–1.1	0.35–7.5	2.0–10.5	20.0	73.3

Table 1. *Cont.*

Building	Monument Registration Number	Period	Type	Usage	Condition	Structure Material	Characteristics of Multiple Leaf Masonry Walls: Wall Thickness From	Wall Thickness to	Width of Leaves ext.	Width of Leaves int.	Leaves Ratio int./ext.	Slenderness Ration (λ)	Core Ratio %	Core Ratio int./W.
Wakala al-Sultan Qaytbay	75		Commercial	Wikala	Partially collapsed	Limestone, lime-based mortar, rubble infill	0.7	1.7	0.20–0.25	0.3–1.2	1.5–5	2.5–6.5	42.9	70.6
Complex of Sultan al-Ashraf Barsbay	121		Religious, Educational, Funerary	Khanqah, Madrasa, Mausoleum	Ruins	Limestone, lime-based mortar, rubble infill	0.6	1.8	0.20–0.30	0.2–1.4	1–7	6.0–8.0	33.3	77.8
Complex of Sultan al-Ashraf Barsbay	175		Religious, Educational, Funerary, Waterworks	Masjid, Madrasa, Mausoleum, Sabil	Fair	Limestone, lime-based mortar, rubble infill	0.9	3.9	0.20–0.25	0.5–3.4	2–13.5	3.5–13.0	55.6	87.2
Complex of Sultan al-Ashraf Inal	158		Religious, Educational, Funerary	Masjid, Madrasa, Mausoleum	Partially collapsed	Limestone, lime-based mortar, rubble infill	0.4	1.6	0.20–0.25	0.08–1.2	0.5–6.0	6.0–10.0	25.0	75.0
The aqueduct (Majra Al-'Uyun)	78		Waterworks	Aqueduct	Fair	Limestone, lime-based mortar, rubble infill	2.9	3	0.25–0.35	2.3–2.5	7.5–10	4.5–6.2	79.3	83.3
Bab Qaytbay	278		Military	Gate	Partially collapsed	Limestone, lime-based mortar, rubble infill	0.7	6.75	0.20–0.30	0.3–6.25	12.5–25	1.0–4.0	42.9	92.6
Masjid Bardbak (Umm al-Ghulam)	25		Religious	Masjid	Fair	Limestone, lime-based mortar, rubble infill	0.6	1.1	0.20–0.25	0.2–0.7	1–3.5	8.0–10.0	33.3	63.6
Complex of Sultan Qansuh al-Ghuri	65–67		Religious, Educational, Funerary	Masjid, Madrasa, Mausoleum	V. good	Limestone, bricks, lime-based mortar, rubble infill	0.8	2	0.20–0.25	0.4–1.4	2–4.5	8.0–17.5	50.0	70.0

Table 1. *Cont.*

Building	Monument Registration Number	Period	Type	Usage	Condition	Structure Material	Characteristics of Multiple Leaf Masonry Walls							
							Wall Thickness		Width of Leaves		Leaves Ratio	Slenderness Ration (λ)	Core Ratio %	
							From	to	ext.	int.	int./ext.			int./W.
Complex of Sultan Qaytbay	104		Educational, Funerary, Waterworks	Madrasa, Mausoleum, Sabil	Ruins	Limestone, lime-based mortar, rubble infill	0.6	1.6	0.20–0.25	0.2–1.1	1–4.5	6.0–13.5	33.3	68.8
Aydakin al-Bunduqdari	146		Funerary, Religious	Mausoleum, Khanqah	Poor	Limestone, lime-based mortar, rubble infill	1.2	2.8	0.20–0.25	0.8–2.3	4–9.2	5.5–8.5	66.7	82.1
Complex of Sultan al-Mu'ayyad Shaykh	190		Religious, Educational, Funerary	Masjid, Madrasa, Mausoleum	Good	Limestone, lime-based mortar, rubble infill	1.9	4.5	0.20–0.25	1.5–4	7.5–18	5.2–8.5	79.0	88.9
Sabil Ibrahim Agha Mustahfizan	238	Ottoman (AD 1517–1867)	Waterworks	Sabil	Fair	Limestone, bricks, lime-based mortar, Gypsum mortar, rubble infill	0.6	1.5	0.15–0.20	0.2–1	1–4	3.0–10.0	33.3	66.7
Bab al-'Azab	555		Military	Tower	V. good	Limestone, lime-based mortar, rubble infill	1.2	2.5	0.22–0.25	0.7–2	3.2–8	6.5–11.5	58.3	80.0
Bab of Ahmad Katkhuda	U93		Military	Tower	Good	Limestone, lime-based mortar, rubble infill	0.6	2.1	0.20–0.25	0.2–1.6	1–6.5	3.5–7.5	33.3	76.2
Saray Al-Musafir khana	20		Palatial	Palace	Ruins	Limestone, bricks, lime-based mortar, Gypsum mortar, rubble infill	0.6	1.5	0.15–0.25	0.3–1.1	1–5.5	6.0–12.0	50.0	73.3
Takiyya Ibrahim al-Kulshani	332		Religious, Funerary	Mausoleum, Khanqah	Ruins	Limestone, lime-based mortar, rubble infill	0.6	1.5	0.15–0.25	0.3–1	1–4	8.0–16.0	50.0	66.7
Zawiya of 'All al-Maghrabi	U4	AD 1866	Religious	Zawiya	Ruins	Limestone, lime-based mortar, rubble infill	0.8	1.1	0.15–0.20	0.5–0.8	2–2.5	6.0–11.5	62.5	72.7

2. Field Survey

2.1. Adopted Methodology

An intensive survey was carried out for each building. The fieldwork focused on construction technology, building materials, and dimensions of multiple-leaf stone masonry walls. The survey included the overall thickness of the walls and their construction materials and the relative thickness of each layer, the overall slenderness ratio of the walls, block dimensions, the composition of the inner core layer, and the connectivity between the walls' leaves.

In the meantime, the constituent materials (stones, bricks, mortars) of the selected buildings were characterized and laboratory tested. The survey included various typologies and usages of historic masonry buildings in order to correctly investigate the construction criteria of multi-leaf stone masonry walls and covered different historical periods from the *Fatimid* to the *Ottoman* era.

The methodology adopted for the fieldwork allowed characterizing the transversal section of the multiple-leaf masonry walls and also yielded definitions of some critical parameters regarding the structural typology of the multiple-leaf masonry walls, i.e.,:

(i) The relative thicknesses of the different layers;
(ii) The ratio between the thicknesses of the inner layers and the total cross-section of the wall;
(iii) The filling materials used to construct the inner core layer;
(iv) The connectivity/constraints types between the inner and external layers;
(v) The construction techniques of the multiple-leaf masonry wall.

Furthermore, three hypotheses were established concerning the utilized construction techniques through extensive field and laboratory investigations.

The adopted geometrical survey followed the methodology provided by Binda et al. [43], where visual inspection and graphic and photographic procedures were conducted to reproduce the metric survey of the studied walls by approximating the geometry with analytical expressions to calculate the percentages of voids, stone units, and mortar joints. Various stone, mortar, and core infill samples were collected and laboratory investigated, and the results were reported in detail in [7,44]. Since the behavior of masonry walls is highly dependent on the technique of construction, the obtained geometrical parameters —together with the characterization of chemical, physical, thermal, petrographic, and mechanical properties [7,25,44] of the materials—give the possibility of better describing multiple-leaf masonry walls and constitute a fundamental basis for the mathematical modelling and the appropriate choice of conservation and retrofitting techniques.

2.2. Major Findings

2.2.1. Usage

To a large extent, the multiple-leaf stone masonry walls were used primarily in the lower floor levels and could extend to the entire wall height of external façades or some major interior walls of the building. In addition, most of the thick stone masonry walls were built according to the multi-leaf system with two outer leaves of more robust material and an inner one of weaker materials; this allowed for a thicker wall cross-section and consequently higher own weight (which increases the overall stability of the wall) at a comparative economic cost.

2.2.2. Geometry

The majority of the multi-leaf walls have an overall thickness greater than 0.5 m, ranging from slender to monolithic walls with a variable relative thickness of inner and external layers. Figure 3 provides typical cross sections of multiple-leaf stone masonry walls of different thicknesses. Mostly, the external layers were built with a homogeneous distribution of the regular well-dressed stone blocks, with nearly constant or uniform dimensions (ashlar facing system). Less homogeneously distributed stone rubble units randomly cut from quarries were often used as infill between the outer leaves. In some cases, a transversal connection between the inner and external skins was provided by irregularly disposed and through passing elements. These transversal bond elements were used to enhance the connectivity between the wall's leaves.

Figure 3. Typical cross-sections of different thicknesses of multiple-leaf stone masonry walls: (**a**) slender wall [$\frac{Int.}{Ext.} \leq 0.35$], (**b**) wall of medium thickness [$0.35 < \frac{Int.}{Ext.} < 1.5$], (**c**) thick (monolithic/massive) wall [$\frac{Int.}{Ext.} \geq 1.5$].

According to the collected data and field survey results, the thickness of the three-leaf stone masonry walls ranged from 0.5 to more than 6 m (Figure 4). It should be stressed that walls with thickness lower than 0.5 m were rare (about 0.94% of the total surveyed walls), while the percentage of the walls with a thickness of more than 4m was about 6.6% of the total surveyed walls. Subsequently, the field survey results confirmed that the majority of the multi-leaf stone masonry walls had a thickness range from 0.5 to 4.0 m.

It is worth mentioning that the walls that were more than 4.0 m thick, and may have reached more than 6 m, were commonly used in the high-rise main facades (as in the Complex of *Al-Sultan Hasan*) and were often used in military defense buildings (as in *Bab Qaytbay*).

This high value of the walls' thickness could be attributed to the need to increase the wall's strength and stiffness in order to enhance the overall mechanical stability and safely resist lateral loads and any expected impact loads. On the other hand, buildings with residential or religious utilization were commonly built with a wall thickness of fewer than 4 m. Thus, it can be inferred that the thickness of the multi-leaf masonry walls was mainly dependent on the function of the building and the overall height of the wall.

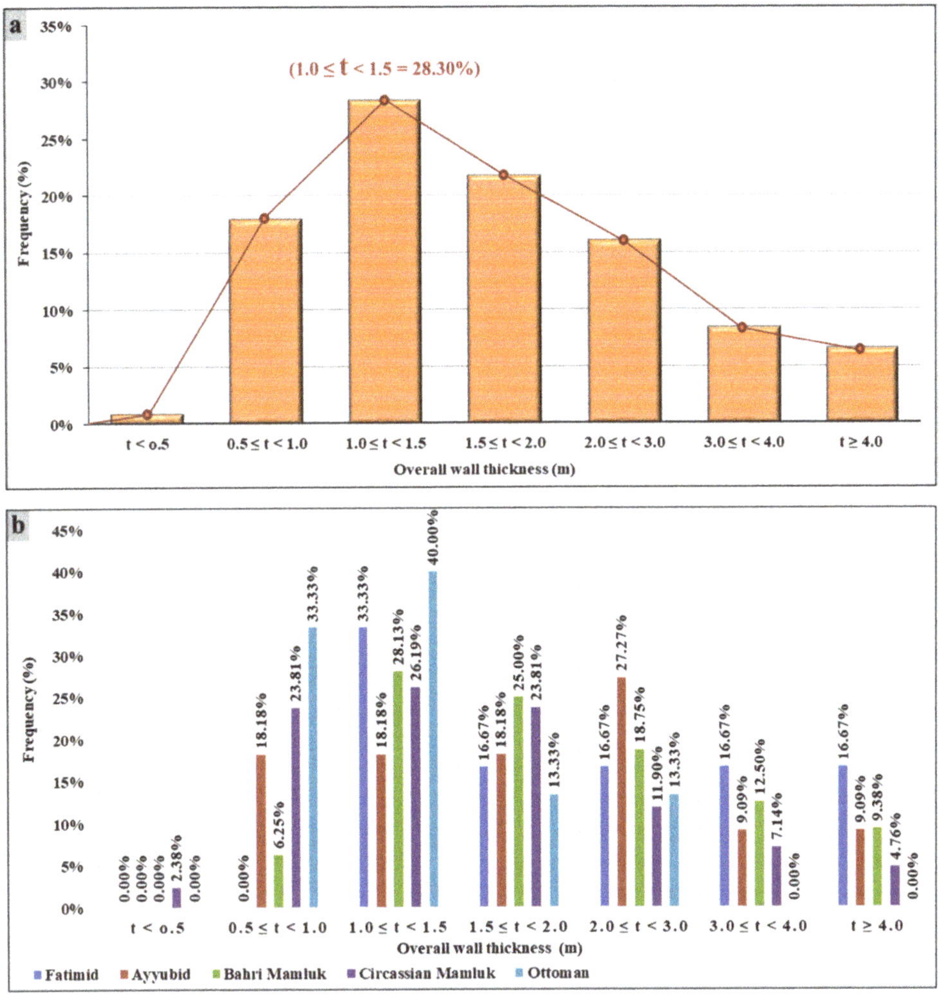

Figure 4. Analyses of the field survey results regarding the thickness of the multiple-leaf masonry walls: (**a**) all surveyed buildings; (**b**) surveyed buildings according to their historic period; *Fatimid, Ayyubid, Bahri Mamluk, Circassian mamluk* and *Ottoman* periods, respectively.

2.2.3. Typology

Three main typologies could be characterized from the analysis of collected data and field survey results as follows:

1. Walls with straight collar joints, where the inner layer was not interconnected with the outer layer. Usually, the inner layer was constructed with an utterly cohesionless infill held by two separate external leaves, while the outer leaves were made of well-dressed limestone blocks with nearly uniform dimensions, ashlars limestone facing system, bonded in horizontal courses. Only a thin interface layer of mortar was found between the three layers (see Figure 5).
2. Walls with keyed collar joints, where the external layer was built with ashlars limestone blocks with variable thickness, creating shear studs to ensure proper connectivity with the inner core infill (see Figure 6a–d).
3. Walls with the transversal bond or tie elements in the cross-linking of the three layers by the use of either header stones (through stones), timber logs or circular marble elements; this ultimately ensured the composite behavior of the wall and enhanced the global performance with the applied external loads (see Figure 6e,f).

Figure 5. Examples of multi-leaf walls with straight collar joints and a cohesionless inner core layer built of rubble stone masonry: (**a**) *Khanqah of Aydakin*, (**b**) *Complex of Al-Sultan Hasan*, (**c**) *Zawiya of 'All al-Maghrabi*, (**d**) *Khanqat Khawand Umm Anuk*, (**e**) *Complex of Al- Sultan Qalawun*, and (**f**) Wakala *al-Sultan Qaytbay*.

Figure 6. Transversal bond elements between the inner and external leaves of the wall: (**a–d**) cross-header stones/shear keys (Complex of Sultan *al-Ashraf Barsbay*, *Amir Alin Aq Palace*, *Bab Qaytbay*, and *Qasr al-Amir Qawsun*, respectively); (**e**,**f**) marble circular cross-joists, (*Masjid al-Salih Tala'i'* and *Bab Zuwayla*, respectively). Adapted from [7].

2.2.4. Constituent Materials

According to the conducted metric survey regarding the composition of multi-leaf stone masonry walls, stone constituted at least 60% of the section, while mortar constituted 25% to 37%. Limestone was the most used stone unit in both external and inner layers. The maximum dimension of rubble stones in the core layer varied from 10 to about 40 cm, with an average of 20 cm, while the average stone block in the outer leaves measured 25 × 35 × 50 cm. Most of the mortars were made of air-hardening lime, usually with a powdery and brittle consistency [7,25], and the dimensions of mortar joints were irregular and, on average, varied between 1 and 5 cm. Mostly, the core layer of the majority of the survey walls was built mainly of stone-rubble infill. Nevertheless, in some buildings dating back to the Ottoman period—for example *Sabil Ibrahim Agha Mustahfizan* and *Saray Al-Musafir khana* (see Figure 7)—the core layer was built of brick rubble.

Figure 7. Different core infill materials of multiple-leaf masonry walls: (**a–c**) stone-rubble infill—Complex of *Sultan al-Ashraf Barsbay*, (**d**,**e**) brick rubble infill, (**a**,**b**) Complex of *Sultan al-Ashraf Barsbay*, (**c**) *Khanqat Khawand Umm Anuk*, (**d**) *Sabil Ibrahim Agha Mustahfizan*, (**e**) *Saray Al-Musafir khana Zawiya of 'All al-Maghrabi*.

2.2.5. Characterization of the Inner and External Layers

The thickness of both external layers was mostly similar and usually less than the inner layer. Also, the percentage of their thickness in relation to the overall thickness of the wall was nearly constant. Furthermore, these external layers were mainly built of well-dressed stone blocks with nearly uniform dimensions (ashlars limestone facing system). The dimensions of the stone-rubble infill varied according to the total thickness of the inner layer. In the case of a thick to monolithic core layer, the thicknesses of mortar joints were usually much lower than the thickness of stone rubble, and the percentage of voids was relatively low; consequently, the presence of rubble stones settled in a high percentage ensured a dense and well compacted core infill.

The thickness of the core layer of the majority of the surveyed walls ranged from 0.5 to 2 m, and the percentage of the core layer with thickness lower than 0.5 m was about 14.5% (see Figure 8).

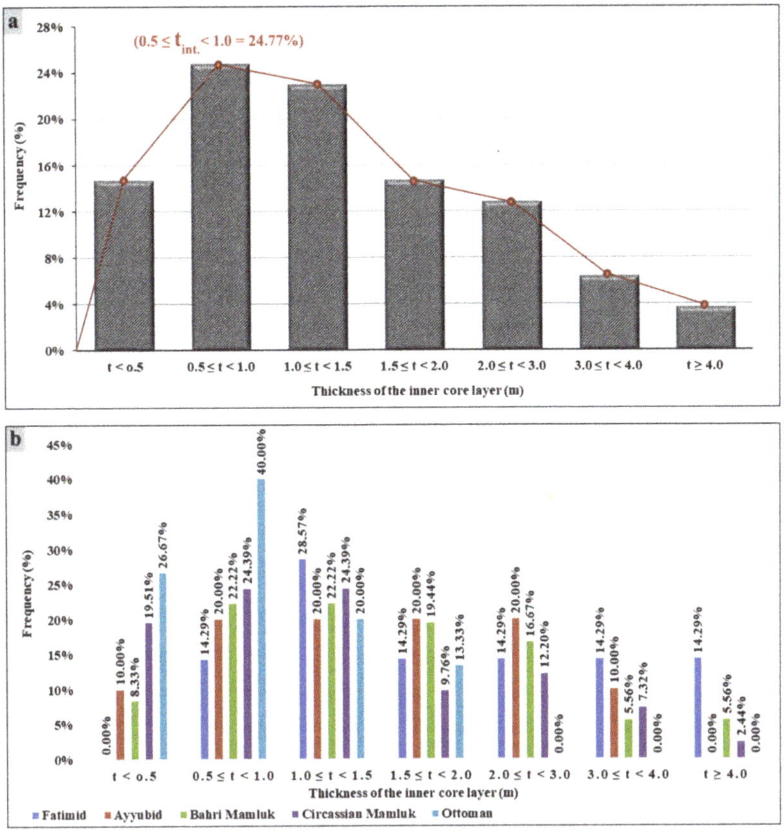

Figure 8. Analyses of the field survey results regarding the thickness of the inner core layer of the multiple-leaf masonry walls: (**a**) all surveyed buildings; (**b**) surveyed buildings according to their historic period: Fatimid, *Ayyubid*, *Bahri Mamluk*, *Circassian mamluk*, and *Ottoman* periods, respectively.

2.2.6. Relative Thickness of the Layers

Based on the analysis of fieldwork results, it can be stated that the ratio between the thicknesses of inner and outer layers depended entirely on the required thickness of the wall. Since the thickness of both external layers was similar and also their percentage to the overall thickness of the wall was about constant, the ratio between the core layer's thickness and the overall thickness of the wall varied according to the total thickness of the cross-section, see Figure 9. Moreover, the ratio between the thicknesses of inner and exterior layers (int./ext.) in most of the walls ranged from 1 to 10, where this percentage increased as the total thickness of the wall increased.

In the case of the cohesive core infill layers that were characterized by good mechanical properties and relatively high stiffness, the thickness of the core layer was considerably larger than the thickness of the outer layer. This could be observed—for example—in high-rise and military defense constructions, where the total wall thickness exceeded 4 m. In this case, a much higher proportion of the applied loads was countered by the inner layer. Since the applied vertical compressive load on the wall is distributed according to the ratio between modulus of elasticity and thickness of the wall, assuming uniform vertical strain (as shown by Egermann et al. [45] and Binda et al. [8,46]), a simplified modeling of the distributed vertical loads, P_e and P_i, acting on the outer and inner leaves, respectively, could be obtained from Equation (1).

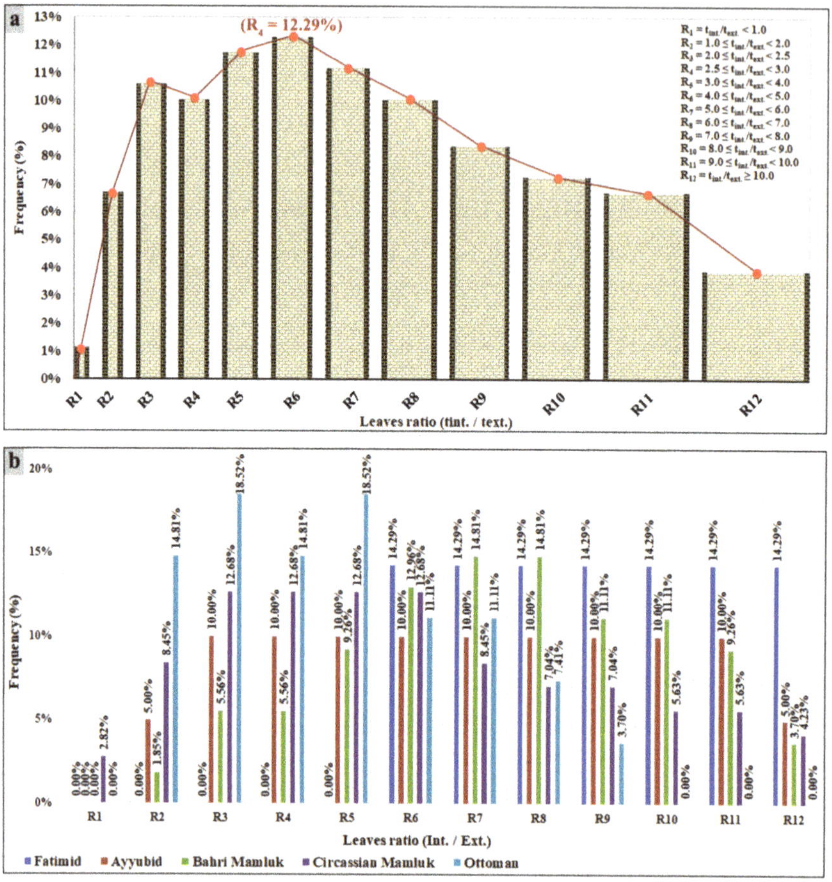

Figure 9. Analyses of the field survey results regarding the ratio between the internal and external leaves thicknesses of multiple-leaf masonry walls: (**a**) all surveyed buildings; (**b**) surveyed buildings according to their historic period: *Fatimid, Ayyubid, Bahri Mamluk, Circassian mamluk* and *Ottoman* periods, respectively.

$$P_i = \left(\frac{1}{1 + 2\frac{E_e t_e}{E_i t_i}}\right) p \quad P_e = \left(\frac{1}{2 + \frac{E_i t_i}{E_e t_e}}\right) p \tag{1}$$

where p is the initial total applied load; E_e and E_i are the moduli of elasticity of the outer and inner leaves respectively; t_e and t_i are the thicknesses of outer and inner leaves, respectively. Further details of the load transfer mechanism will be provided in a forthcoming paper.

Under the same characteristics of the wall, loading, and boundary conditions, the increase in the thickness of core layer and, consequently, the increase in the overall thickness of the wall, significantly reduced the resultant internal compressive and shear stresses.

Furthermore, the thickness of the walls often increased as the total height increased, keeping the slenderness ratio between 3 to 12 (Figure 10). However, a relatively small number of the surveyed walls had a ratio larger than 12.

Finally, the obtained results of the fieldwork alongside the analysis of the typical cross-sections of the three-leaf masonry walls and the survey of different typologies have contributed to form a thorough scientific base for the construction hypothesis of this distinctive type of historic wall, which will form a fundamental reference for any further experimental studies on their structural behavior.

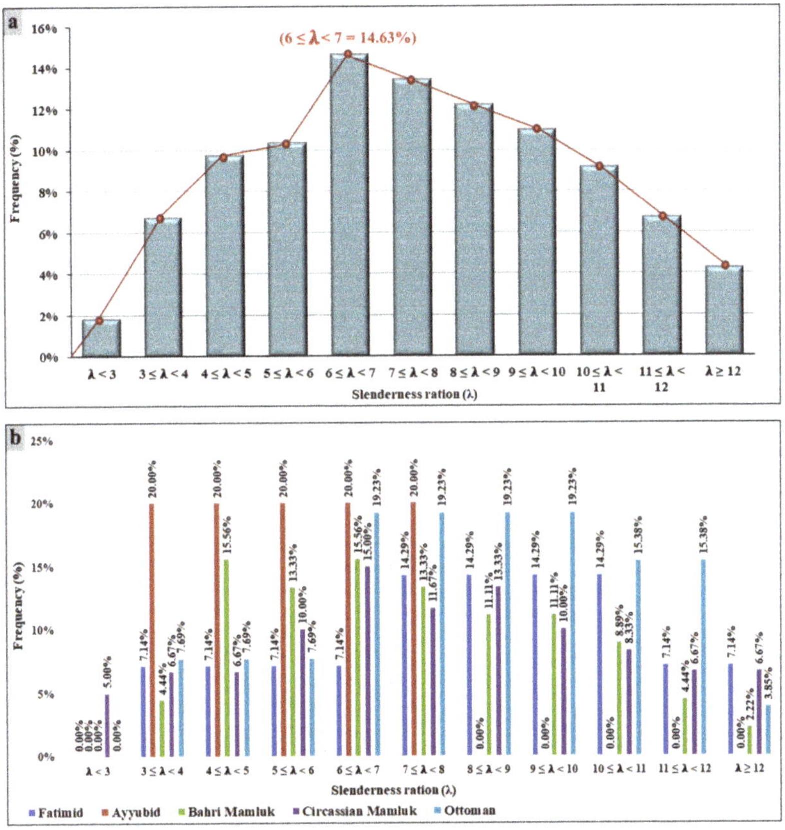

Figure 10. Analyses of the field survey results regarding the slenderness ratio of the multiple-leaf masonry walls: (**a**) all surveyed buildings; (**b**) surveyed buildings according to their historic period: *Fatimid*, *Ayyubid*, *Bahri Mamluk*, *Circassian mamluk* and *Ottoman* periods, respectively.

3. Construction Hypotheses of Multiple-Leaf Masonry Walls

This section focuses on the construction hypotheses of multi-leaf stone masonry walls in accordance with the field survey results. Although multi-leaf stone masonry walls are one of the common building systems and have been widely used in a variety of historical constructions, there is a considerable lack of information about the utilized methods and techniques for their construction. This type of wall can be characterized by different construction techniques, which essentially affect their mechanical behavior, in particular the interaction between the layers and the overall integrity of the wall.

Three hypotheses could be made on the construction techniques based on the typology, overall thickness of the wall, and the connectivity between the inner and outer layers, as described in the following subsections.

3.1. Construction Hypothesis (I): Weak Inner Core

A first hypothesis assumes that the multi-leaf masonry walls were built with almost a thoroughly segregated/cohesionless internal core retained by two separate external leaves (Figure 5). The walls were mainly constructed by building the outer two layers first, leaving a space or cavity of the designed thickness for the core infill between the outer leaves by laying a course of stretcher stone blocks along the chalk line for the entire length of the wall after determining the opening positions (Figure 11a). Then, in correspondence with the corners and openings, adjacent courses of stone blocks were carefully laid. These provided a guide for laying the remaining blocks, so they should be accurately leveled and correctly spaced. String lines were then stretched between corners of the wall or between a corner and an opening's edges (Figure 11b). To ensure that the wall would be straight and not bowed or curved, the first course of the lead was carefully aligned and was usually at least four or five units long. Leads were usually built four to five courses higher than the central part of the wall, and each course of the lead was carefully checked to accurately assure that it was leveled in both directions and plumb.

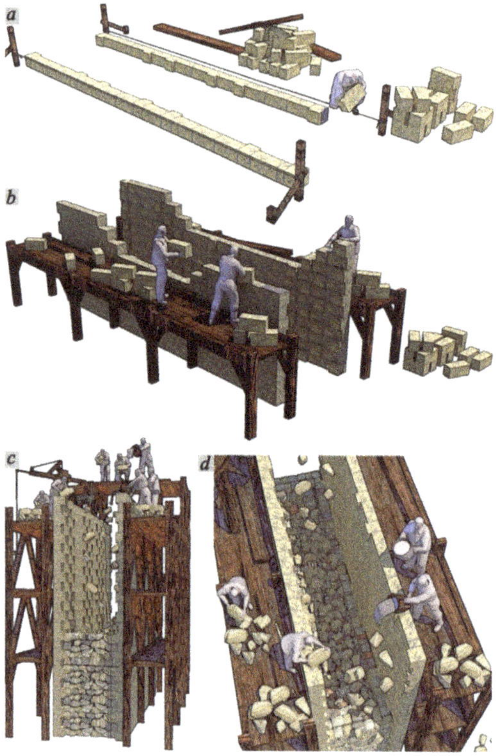

Figure 11. Construction hypothesis (I): (**a**) laying the first course of stretcher stone blocks along the entire length of the wall; (**b**) construction of the two outer leaves; (**c**,**d**) pouring and throwing of the inner core leaf.

After that, the outer shells were brought up to the total height of the wall by laying the stone blocks up with mortar in specific patterns, providing a typical running bond in which one unit overlapped the unit in the course below by almost half its length. This bonding was aimed at homogenously spreading the loads and increasing the interconnectivity between the stone blocks. In most cases, timber logs were embedded horizontally along the length of the wall at considerable interval heights (about 1.0 to1.50 m), as shown in Figure 12, in order to properly adjust the wall verticality, ensure that walls were going up plumb, and offer a good bedding level for successive wall courses. When the two outer layers had been brought up to the required height of the wall, the core infill was simply poured or thrown until the top of the wall was reached (Figure 11c,d). This infill was generally composed of rubble stones bonded with mortar. In some cases, the three layers may be bonded together using transversally bound elements of stone, wood, or marble. However, their use may have impeded the casting process of the inner layer.

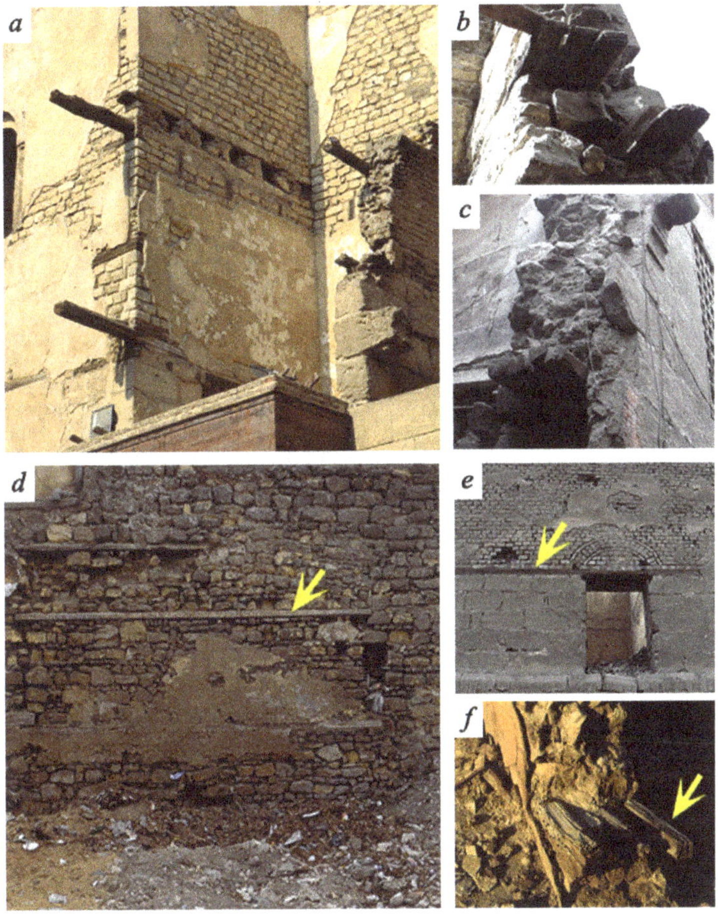

Figure 12. Timber tie system in stone masonry constructions: (**a**) Amir Alin Aq Palace, (**b**) Sabil Ibrahim Agha Mustahfizan, (**c**) Qasr al-Amir Qawsun, (**d**,**e**) Qasr al-Amir Qawsun, (**f**) Takiyya Ibrahim al-Kulshani.

It is believed that, during construction, the outer leaves were temporarily braced with the use of timber cross-braces frame (timber ledger) so that the pouring of the core infill was suitably set to prevent any lateral loads produced during the casting/filling process from exceeding their structural capabilities and to avoid detachment, buckling, or expulsion of the outer leaves.

The construction of multi-leaf walls by following this technique produces walls with a weak and incoherent core layer. As a result of casting the core infill after the complete construction of the outer layers, segregation may occur between the mortar and the rubble infill. Accordingly, it is believed that this method was of limited use for thin walls and walls of limited heights. Nevertheless, it is interesting to observe that this method has been exploited in more recent historic buildings, as in the construction of *Gebel Aulia* dam in Sudan, 1904; *Mohamed Aly* barrage in Egypt, 1941; and Aswan High Dam, 1970, where the well-dressed granite blocks were used as facing/shuttering layers, while granite rubble masonry was used for filling the cross-section (see Figure 13). Through the development of the construction materials and for structural upgrading purposes, the rubble masonry of the core layer was replaced with mass concrete and the timber logs were also replaced by reinforcing steel bars, as in the majority of the old irrigation structures in Egypt [47].

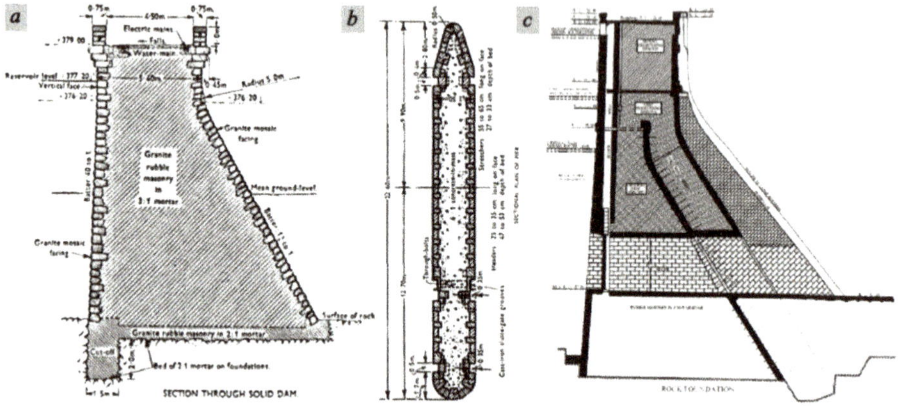

Figure 13. Archived cross-sectional drawings of (**a**) *Gebel Aulia* Dam, (**b**) *Mohamed Aly* Barrage, and (**c**) *Aswan* High Dam.

3.2. Construction Hypothesis (II): Monolithic Core Infill

The second hypothesis concerning the construction technique adopted for monolithic multi-leaf masonry walls assumes a thick core infill faced with relatively light sheeting of ashlar facing stone masonry to form the enclosed/composed element. According to this hypothesis, the inner core layer was firstly constructed in the intermediate space of the total cross-section of the wall by piling up undressed, rough stones, as they were randomly cut from quarries and sufficiently bonded with mortar without being committed to horizontal courses. In this case, the longitudinal bond in the core skin was achieved by overlapping stones in roughly adjacent courses, with a variable amount of overlap as the stones varied in size. Relatively large stone blocks were cut in a reasonably prismatic block shape and were used for corners and the jambs of openings to obtain increased strength and stability at these locations. One of the most important concerns regarding the construction of the inner core layer was keeping it straight and plumb; this could be established by building leading sections or leads at the ends or corners of walls and string lines were then stretched between corners. These leads provided a guide for laying the remaining infill mixture at the middle of the wall length, see Figure 14a–c. After that, the inner core layer was completed until the required height of the wall. Then, it was coated/faced with two outer layers of regular well-dressed stone blocks (ashlar stone masonry facing system) (see Figure 14d,e).

Even though the inner layer is characterized by high cohesion and a significant load-bearing capacity, with a low percentage of voids, it was weakly bonded with the outer layers by only a thin interface layer of lime-based mortar, affecting the homogenous performance of the wall. Consequently, the internal leaf resists a relatively large portion of the applied vertical loads, while a small portion is carried by the outer layers. Besides, the adhesion between the leaves would have been scarce or null, and the inner core layer would exert lateral load on both outer leaves, causing detachment and separation. To overcome these deficiencies, the cohesion between the three layers was usually achieved through drilling some holes/apertures at the inner layer, in correspondence with specifically designed positions to insert the transversal bond or tie elements in the cross-linking of the three layers by the use of either header stones (through stones), timber logs, or circular marble elements.

An example of a more recent historic construction that confirmed this construction hypothesis is the Sanayta lock and bridge-regulator construction in Egypt (Figure 15) in 1899: the inner core layer of the multi-leaf walls of the lock were built first of brick masonry with *Homra* mortar, and then faces were cladded with cement mortar. The main portion of the wall has a width diminishing from 3 m to 1.2 m, by steps of 0.25 m. At the upstream end, these walls are encased with ashlar masonry in 0.4 m courses [48].

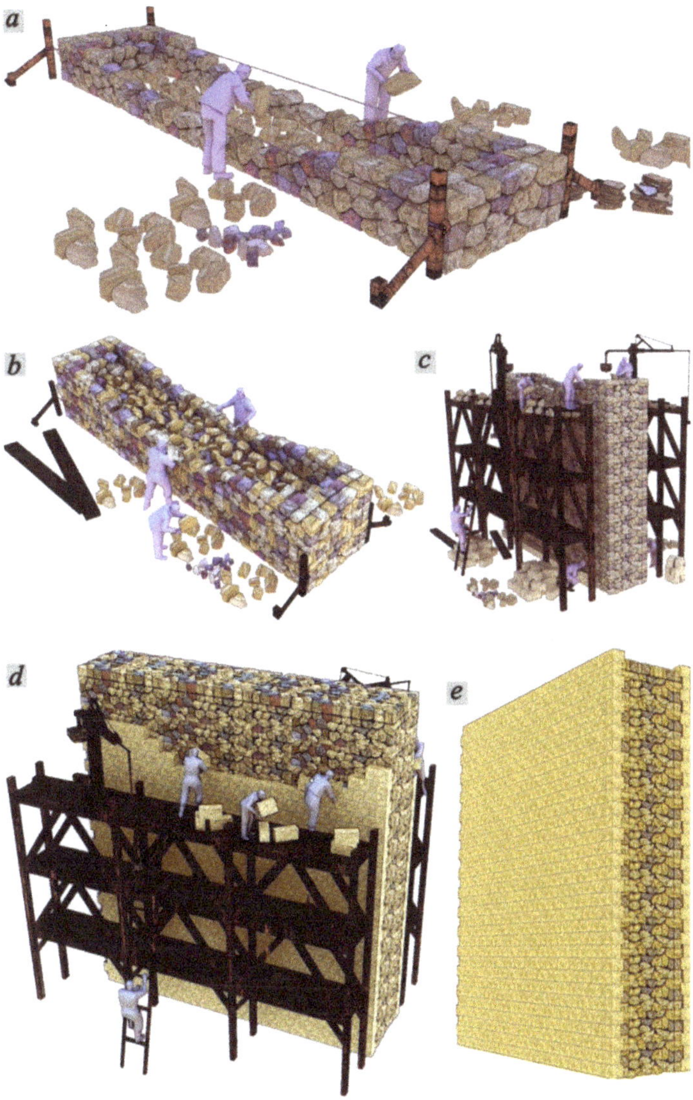

Figure 14. Construction hypothesis (II): (**a**–**c**) construction of the inner core layer by randomly piling up undressed, rough stone, which was bound with mortar without being committed to horizontal courses; (**d**) construction of the two outer leaves with regular well-dressed stone-blocks; (**e**) the full tree-leaf stone masonry wall.

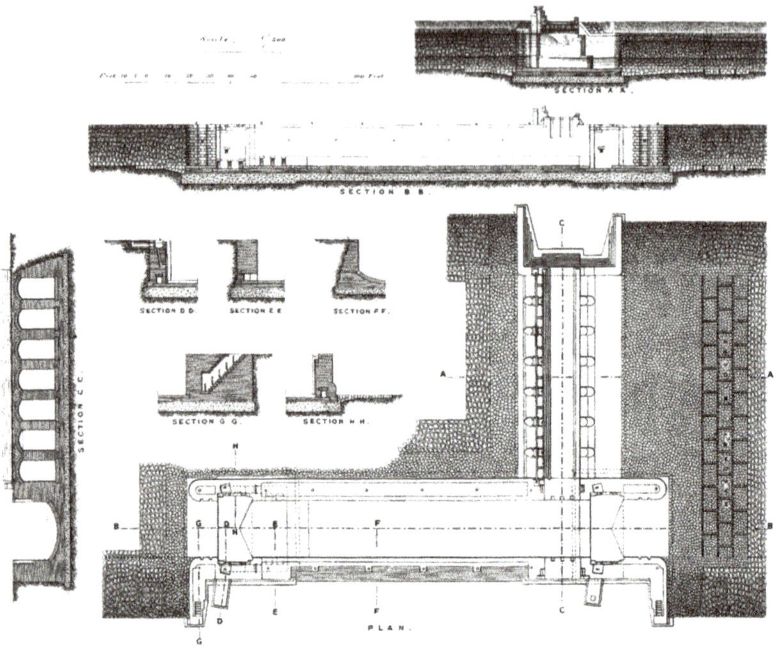

Figure 15. Archived drawings of Sanayta lock and bridge-regulator construction, adapted from [48].

3.3. Construction Hypothesis (III): Thick Core Infill

The third hypothesis on the construction technique adopted in three-leaf stone masonry walls is similar to the second one, since it is assumed that the inner core layer is characterized by high cohesion, good mechanical properties, and an actual load-bearing capacity with low percentages of voids. Furthermore, the presence of transversal bond elements (stone, timber, or marble joists) between the inner and external layers enhances the connectivity between the wall's leaves. The three layers are perfectly homogeneous in this type and behave as a single composite element to resist the applied loads, so that the lateral pressure of the inner core layer thrusting the outer leaves decreases significantly and almost vanishes. The vertically applied loads on the wall cross-section are distributed according to the relative stiffness of the core infill and the outer skins.

As regards this hypothesis, the three layers of the wall were simultaneously built. The two outer layers of the wall were constructed up to a suitable height (presumably three to four courses), and instantaneously the core infill was constructed in between. The two outer leaves were mainly built using well-dressed limestone blocks with nearly uniform dimensions (ashlar facing system), while the inner core layer was constructed mainly by successive layering of rubble stones and mortar bedding, which was not cast but layered more or less accurately by the masons (see Figure 16).

Providing a proper setting time for the infill mixture or mortar, the two outer layers were then continued up to an appropriate height and then the inner layer was also built up to the same height of the wall. Up to a certain height from the ground, the three layers could be constructed from both sides of the wall without any scaffolding (Figure 17a). Subsequently, it was essential to assemble the scaffolding for building the upper parts; the stone blocks were then lifted to the scaffolding platform by a hoist (Figure 17b–d).

Figure 16. Different cross-sections of multiple-leaf stone masonry walls with the internal leaf built by successive layering of rubble stones and mortar bedding with horizontal courses: (**a**) Complex of *Al-Sultan Hasan*, (**b**) Complex of *Sultan al-Ashraf Barsbay*, (**c**) *Qubba al-Amir Tarabay al-Sharifi*, (**d**) and (**f**) perimeter wall of *Bab Qaytbay*; and (**e**) *the aqueduct* (*Majra Al-'Uyun*).

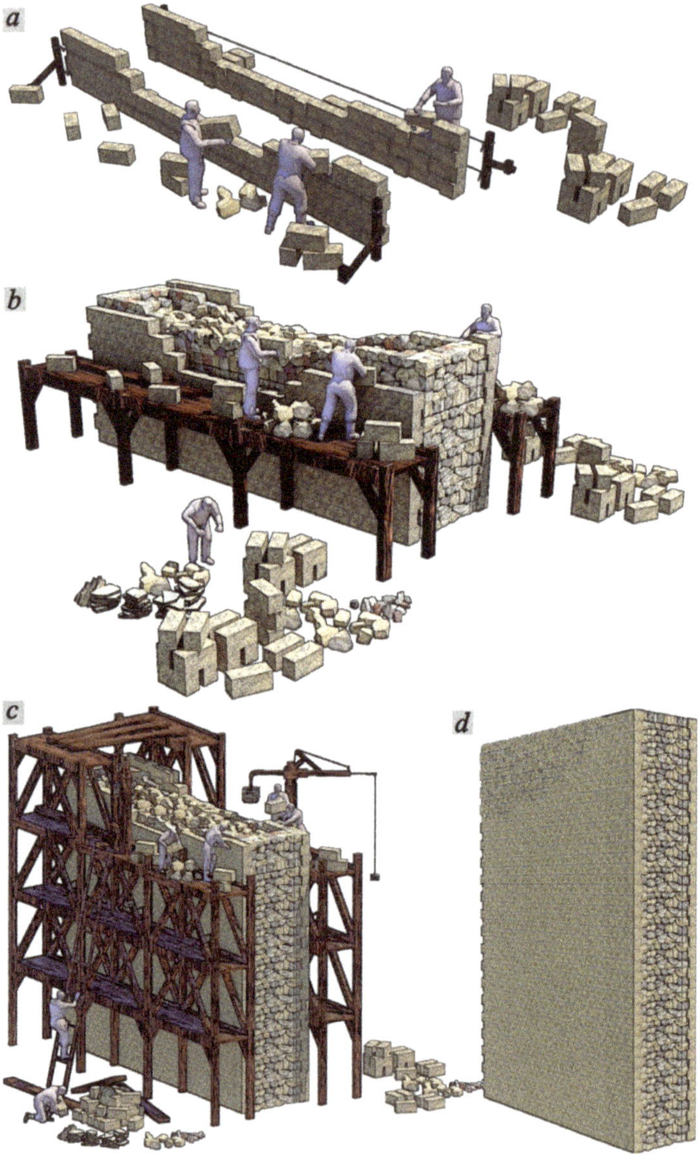

Figure 17. Construction hypothesis (III): (**a**) constructing the outer leaves up to a suitable height; (**b**) constructing the inner core layer by successive layering of rubble stones and mortar bedding with horizontal courses; (**c**) assembly of scaffolding erection for the construction of the upper parts; (**d**) the full tree-leaf stone masonry wall.

4. Discussion

The field study allowed concluding that the multi-leaf stone masonry walls varied in their geometrical characteristics, typologies, and construction techniques; consequently, according to their typology, they are characterized by different structural performance and load-transfer mechanisms. Classification of three types of multi-leaf stone-masonry walls was conducted (see Section 2.2.1) according to the complexity of the wall, the ratio between the thickness of the inner layer to the thickness of the outer layer, the composition of the inner core layer, and the type of bonding between the inner and outer layers.

In the case of a poor/incoherent core layer, the voids were indiscriminately diffused and the percentage of mortar was relatively high, in addition to the poor bonding between this inner layer and the outer layers due to the absence of any transversal bond elements (straight collar joints). Friction was the prevalent resistant mechanism, while cohesion within the internal leaf and adhesion between the leaves was scarce. In this case, the core layer resisted a very small portion of the applied vertical loads as most of the load was carried by the outer shells. Besides, the cohesionless core rubble infill exerted non-linear lateral pressure thrusting the outer leaves.

As a result of constructing the wall by this technique, the inner layer was nearly cohesionless and almost separated from the outer leaves; moreover, an indiscriminate diffusion of voids and relatively high percentages of mortar were formed. As a result, complex crack patterns, detachment, buckling, and expulsion of the outer leaves characterized the collapse mechanism. Accordingly, this construction method was commonly used in the case of a thin multi-leaf wall of low height (funerary, educational, and residential buildings).

In the case of defense structures, waterworks, military, and palatial and commercial buildings, another type of multiple-leaf masonry wall was utilized with an inner core layer characterized by high cohesion, good mechanical properties, and a considerable load-bearing capacity. The inner core layer had a low percentage of voids, and the percentage of mortar was relatively low, while the presence of rubble stones settled at a high percentage, corresponding to relatively high stiffness. In most of the external multi-leaf stone masonry walls, the outer leaves were arranged in a wide variety of ways to produce a satisfactory bond, and each arrangement could be identified by the pattern of headers and stretchers on the face of the wall.

The resistance of monolithic multi-leaf walls, with core infill layers characterized by poor connection between the inner and outer layers, mainly depended on the inner layer. Due to the absence of the transversally bound elements and shear keys, the bond between the leaves was weak, and the inner core layer could exert lateral load on both outer leaves, causing severe damage, detachment, and separation. Nevertheless, in many cases, the inner core layer was still upright, even after the potential collapse of the external layers, and safely resisted the applied vertical load due to its cohesive condition, which confirms that the overall structural behavior of the wall is mainly dependent on the performance of the core infill (see Figure 18).

Figure 18. Examples of detachment and collapse of facing ashlar layers, exposing the inner core layer built of rubble stone masonry: (**a**) *Bab al-Mahruq*, (**b**) *Bab Qaytbay*, and (**c**) northeastern historic Cairo wall.

To achieve effective cohesion between the inner and outer layers and ensure a homogeneous behavior of the wall, the mason usually provided a transversal bond by using long header stones (known as bonders). These transversal elements were extended to sufficient depth through the wall thickness in order to enhance the connectivity between the inner and external leaves and provide lateral reinforcement, which helped to resist lateral loads and overcome the brittle nature of rubble stone masonry. Accordingly, the three layers were able to homogeneously resist vertical and lateral applied loads. The different values of cohesion within the inner core layer and adhesion between the leaves, the possible presence of transverse elements, and different percentages of voids, are all parameters that can lead to relevant differences in the wall's global behavior.

In regards to the above, each type of defined multi-leaf stone masonry wall has its characteristic behavior under the external actions, which mainly depends on the mechanical properties and composition of the inner layer; the mechanical properties, type and arrangement of the stone units of the external layers; and the adopted connectivity between the inner and outer layers. This highlights the significance of considering these parameters when preparing mechanical models to investigate the global performance and detect the load-carrying capacity of this type of historic wall, and also when choosing the material and methods for the intervention and retrofitting. The appropriate numerical models for simulating the mechanical behavior of multi-leaf walls, as well as proposing an appropriate strengthening technique, mainly depend on an accurate investigation aimed at characterizing their geometry and construction techniques, together with the essential physicochemical and mechanical investigations of the utilized building materials.

5. Conclusions

The present research focused on the structural field survey of multiple-leaf stone masonry walls of historic buildings in a variety of historic areas in Egypt. By analyzing the obtained results of this fieldwork together with the available historical data, a general characterization of the common typologies of the transversal section of multiple-leaf stone masonry walls was performed. Moreover, various construction hypotheses of this type of historic construction were proposed, based mainly on the results obtained in the conducted field survey.

The survey results confirmed that the multiple-leaf masonry wall system was commonly used as vertical structural elements in almost all types of complex historical medieval structures in Egypt, i.e., religious, service, residential, fortification, irrigation, etc. These walls present different characteristics depending on their typologies and construction techniques, which gradually changed from multiple-leaf walls with a cohesionless internal core infill of weak mechanical resistance held by two separate external leaves to the fixed rubble-core masonry strongly bonded with the outer leaves, ensuring an overall composite performance of the wall. Usually, transversal tie elements of timber, marble, or stone were used to enhance the connectivity between the inner and external leaves and achieve an integrated composite mechanical performance of the wall.

Three main types of multiple-leaf stone masonry walls were identified according to the thickness and composition of the inner core layer. The monolithic core layer was usually utilized primarily in the historic military defense and hydraulic structures to enhance the overall stability of the wall against any potential lateral or impact loads. On the other hand, walls with relatively thin to thick core layers were used in religious, commercial, and residential buildings.

Since the multi-leaf stone masonry walls vary in their geometrical characteristics, typologies, morphologies, and construction materials, assessing their structural behavior essentially requires a detailed metric survey and comprehensive experimental investigation of the mechanical and physicochemical properties of their components. Additionally, the role of the core layer in carrying the applied external load has to be considered and accurately investigated, together with the prediction of the lateral pressure exerted by the cohesionless core infill on the outer layers. These investigations are a prerequisite for any intended strengthening and conservation practices.

Author Contributions: Conceptualization, O.A., D.A., E.k.M., A.T., and A.S.; methodology, O.A., D.A.; investigation, O.A., A.T., and A.S.; formal analysis, O.A., and E.k.M.; visualization, O.A.; writing—original draft preparation, O.A., and E.k.M.; writing—review and editing, O.A., D.A. All authors have read and agreed to the published version of the manuscript.

Funding: This research received no specific grant from any funding agency in the public, commercial, or not-for-profit sectors.

Data Availability Statement: The data presented in this study are available on request from the corresponding author.

Acknowledgments: We dedicate this article to Ioannis Liritzis, who has significantly contributed to the spread of Egyptian heritage conservation sciences. The authors are also grateful to Nikiforos Meimaroglou, Osama Talat, Eng. Amr Saleh, and Eng. Mustafa Bekhet for the collaboration and significant support.

Conflicts of Interest: The authors declare no conflict of interest.

References

1. Aldreghetti, I.; Baraldi, D.; Boscato, G.; Cecchi, A.; Massaria, L.; Pavlovic, M.; Reccia, E.; Tofani, I. Multi-Leaf Masonry Walls with Full, Damaged and Consolidated Infill: Experimental and Numerical Analyses. *Key Eng. Mater.* **2017**, *747*, 488–495. [CrossRef]
2. Egermann, R.; Frick, B.; Neuwald, C. Analytical and experimental approach to the load bearing behaviour of multiple leaf masonry. In *Structural Repair and Maintenace of Historical Buildings III (STREMA)*; WIT Press: Southampton, UK; Boston, MA, USA, 1993.
3. Drysdale, R.G.; Hamid, A.A.; Baker, L.R. *Masonry Structures: Behavior and Design*; Leon, B.M.D., Zurite, P., Handy, S., Eds.; Prentice Hall: Hoboken, NJ, USA, 1999.
4. Drei, A.; Fontana, A. Influence of geometrical and material properties on multiple-leaf walls behaviour. *WIT Trans. Built Environ.* **2001**, *55*, 2001.
5. Feilden, B.M. *Conservation of Historic Buildings*; International Centre for the Study of the Preservation and the Restoration of Cultural Property (ICCROM): Rome, Italy, 2003.
6. Pulatsu, B.; Bretas, E.M.; Lourenco, P. Discrete element modeling of masonry structures: Validation and application. *Earthq. Struct.* **2016**, *11*, 563–582. [CrossRef]
7. Amer, O.; Aita, D.; Mohamed, E.K.; Torky, A.; Shawky, A. Experimental investigations and microstructural characterization for construction materials of historic multi-leaf stone-Masonry Walls. *Heritage* **2021**, *4*, 135. [CrossRef]
8. Pina-Henriques, J.; Lourenço, P.; Binda, L.; Anzani, A. Testing and modelling of multiple-leaf masonry walls under shear and compression. In *Structural Analysis of Historical Constructions*; Modena, C., Lourenço, P.B., Roca, P., Eds.; Taylor & Francis Group: London, UK, 2004; pp. 299–310. ISBN 04-1536-379-9.
9. Pina-Henriques, J.L. Masonry under Compression: Failure Analysis and Long-Term Effects. Ph.D. Thesis, University of Minho, Braga, Portugal, 2005.
10. Da-Porto, F.; Valluzzi, M.R.; Modena, C. Investigations for the knowledge of multi-Ieaf stone masonry walls. In Proceedings of the First International Congress on Construction History, Madrid, Spain, 20–24 January 2003.
11. De Felice, G. Out-of-Plane Seismic Capacity of Masonry Depending on Wall Section Morphology. *Int. J. Arch. Herit.* **2011**, *5*, 466–482. [CrossRef]
12. Pulatsu, B.; Gencer, F.; Erdogmus, E. Study of the effect of construction techniques on the seismic capacity of ancient dry-joint masonry towers through DEM. *Eur. J. Environ. Civ. Eng.* **2020**, 1–18. [CrossRef]
13. Vintzileou, E.; Tassios, T.P. Three-Leaf Stone Masonry Strengthened by Injecting Cement Grouts. *J. Struct. Eng.* **1995**, *121*, 848–856. [CrossRef]
14. Toumbakari, E.E. Lime-Pozzolan-Cement Grouts and Their Structural Effects on Composite Masonry Walls. Ph.D. Thesis, Katholieke Universiteit Leuven, Leuven, Belgium, 2002.
15. Valluzzi, M.R.; Da Porto, F.; Modena, C. Behavior and modeling of strengthened three-leaf stone masonry walls. *Mater. Struct.* **2004**, *37*, 184–192. [CrossRef]
16. Vintzileou, E. Grouting of Three-Leaf Masonry: Experimental Results and Prediction of Mechanical Properties. 2007, pp. 171–190. Available online: https://www.researchgate.net/publication/242419246_GROUTING_OF_THREE-LEAF_MASONRY_EXPERIMENTAL_RESULTS_AND_PREDICTION_OF_MECHANICAL_PROPERTIES (accessed on 16 August 2021).
17. Vintzileou, E.; Miltiadou-Fezans, A. Mechanical properties of three-leaf stone masonry grouted with ternary or hydraulic lime-based grouts. *Eng. Struct.* **2008**, *30*, 2265–2276. [CrossRef]
18. Mazzon, N. Influence of Grout Injection on the Dynamic Behaviour of Stone Masonry Buildings. Ph.D. Thesis, University of Padova, Padova, Italy, 2010.

19. Valluzzi, M.R.; da Porto, F.; Modena, C. Behaviour of multi-leaf stone masonry walls strengthened by different intervention techniques. In *Historical Constructions: Possibilities of Numerical and Experimental Techniques*; Lourenço, P., Roca, P., Eds.; University of Minho: Guimarães, Portugal, 2001; pp. 1023–1032.
20. Anzani, A.; Binda, L.; Fontana, A.; Henriques, J.P. An experimental investigation on multiple-leaf stone masonry. In Proceedings of the 13th International Brick and Block Masonry Conference, Amsterdam, The Netherlands, 4–7 July 2004.
21. Toumbakari, E.; Gemert, D.V.; Tassios, T.; Vintzileou, E. Experimental investigation and analytical modeling of the effect of injection grouts on the structural behaviour of three-leaf masonry walls. In *Structural Analysis of Historical Constructions*; Modena, C., Lourenço, P.B., Roca, P., Eds.; Taylor & Francis Group: London, UK, 2005; pp. 707–717.
22. Valluzzi, M.R.; Mazzon, N.; Munari, M.; Casarin, F.; Modena, C. Effectiveness of Injections Evaluated by Sonic Tests on Reduced Scale Multi-Leaf Masonry Building Subjected to Seismic Actions. 2009, pp. 2–7. Available online: https://www.researchgate.net/publication/237761734_Effectiveness_of_injections_evaluated_by_sonic_tests_on_reduced_scale_multi-leaf_masonry_building_subjected_to_seismic_actions (accessed on 16 August 2021).
23. Van Gemert, D.; Ignoul, S.; Brosens, K.; Toumbakari, E.-E. Consolidation and Strengthening of Historical Masonry by Means of Mineral Grouts: Grout Development. *Restor. Build. Monum.* **2015**, *21*, 29–45. [CrossRef]
24. Bhandari, S.; Shrestha, J.K.; Pradhan, S. In-plane capacity of multi-leaf stone masonry walls. In Proceedings of the IOE Graduate Conference, Kirtipur, Nepal, May 2019.
25. Amer, O.; Abdel-Aty, Y.; Abdel-Hady, M.; Aita, D.; Torky, A.; Hussein, Y. Multiscientific-based approach to diagnosis and characterization of historic stone-masonry walls: The mausoleum of al-imam al-shafi'i, cairo (egypt). *Mediterr. Archaeol. Archaeom.* **2020**, *20*, 67–82.
26. Creswell, K. *The Muslim Architecture of Egypt*; Hacker Art Books; Clarendon Press: New York, NY, USA, 1978; Volume 1.
27. Behrens-Abouseif, D. Topography and Architecture of the North-Eastern Suburbs of Cairo in the Circassian Mamluk Period. Master's Thesis, American University in Cairo (A.U.C.), New Cairo, Egypt, 1980.
28. Behrens-Abouseif, D.; Fernandes, L. Sufi Architecture in Early Ottoman Cairo. In *Annales Islamologiques*; Institut Français D'archéologie Orientale: Cairo, Egypt, 1984; Volume 20, pp. 103–114.
29. Behrens-Abouseif, D. Change in Function and Form of Mamluk Religious Institutions. In *Annales Islamologiques*; Institut Français D'archéologie Orientale: Cairo, Egypt, 1985; Volume 21, pp. 73–93.
30. Allen, T. *A Classical Revival in Islamic Architecture*; Ludwig Reichert: Wiesbaden, Germany, 1986.
31. Hoag, J. *Islamic Architecture*; Rizzoli: New York, NY, USA, 1987.
32. Behrens-Abouseif, D. The Takiyyat Ibrahim al-Kulshani in Cairo. In *Muqarnas*; Brill: Leiden, The Netherlands, 1988; Volume 5, pp. 43–60.
33. Behrens-Abouseif, D. *Islamic Architecture in Cairo: An Introduction*; American University in Cairo (A.U.C.) Press: Cairo, Egypt, 1989.
34. OICC. *Organization of Islamic Capitals and Cities, Principles of Architectural Design and Urban Planning During Different Islamic Eras: Analytical Study for Cairo City, Jeddah, Saudi Arabia: Markaz al-Dirāsāt al-Takhṭīṭīyah wa-al-Mi'Mārīyah*; Center for Revival of Islamic Architectural Heritage: Cairo, Egypt, 1992.
35. Mal, H. *Domes in Islamic Architecture "al-Quba al-Madfan", Its Evolution Till the end of Mamluk Period*, 1st ed.; Al-Thaqafa al-Deenya Liberary: Cairo, Egypt, 1993. (In Arabic)
36. Jarrar, S.; Riedlmayer, A.; Jeffrey, B.S. *Resources for the Study of Islamic Architecture*; Aga Khan Program for Islamic Architecture: Cambridge, MA, USA, 1994.
37. Hillenbrand, R. *Islamic Architecture: Form, Function and Meaning*; Edinburgh University Press: Edinburgh, UK, 1994.
38. Antoniou, J. *Historic Cairo A Walk through the Islamic City*; Third Printing; The American University in Cairo Press (A.U.C): Cairo, Egypt, 2002.
39. Williams, C. Islamic Monuments in Cairo. In *The Practical Guide*, 5th ed.; The American University in Cairo Press (A.U.C): Cairo, Egypt, 2002.
40. Warner, N. *The Monuments of Historic Cairo, a Map and Descriptive Catalogue, 132*; American University in Cairo Press (A.U.C): Cairo, Egypt, 2005.
41. Yeomans, R. *The Art and Architecture of Islamic Cairo*, 1st ed.; Garnet Publishing: Cairo, Egypt, 2006.
42. Behrens-Abouseif, D. *Cairo of the Mamluks: A History of Architecture and Its Culture*; I.B. Tauris: London, UK, 2007.
43. Binda, L.; Cardani, G.; Saisi, A. A classification of structures and masonries for the adequate choice of repair. In *International RILEM Workshop on Repair Mortars for Historic Masonry*; RILEM, International Union of Laboratories and Experts in Construction Materials, Systems and Structures: Delft, The Netherlands, 2005; pp. 20–34.
44. Amer, O. Experimental and Analytical Studies on Structural Behavior of Multiple-Leaf Masonry Walls under Loads and Biological Factors, and the Appropriate Restoration Techniques with Application on Chosen Historical Islamic Buildings Biologically Affec. Master's Thesis, Conservation Department, Faculty of Archaeology, Cairo University, Cairo, Egypt, 2018.
45. Egermann, R.; Neuwald-Burg, C. Assessment of the load bearing behaviour of historic multiple leaf masonry walls. In Proceedings of the 10th International Brick/Block Masonry Conference (IBMAC), Calgary, AB, Canada, 5–7 July 1994; pp. 1603–1612.
46. Binda, L.; Fontana, A.; Anti, L. Load transfer in multiple leaf masonry walls. In Proceedings of the 9th International Brick/Block Masonry Conference, Berlin, Germany, 13–16 October 1991.

47. Mohamed, E.K.; Khalil, E. Innovative solution for the repair of hydraulic structures (regulators). *Water Sci.* **2018**, *32*, 179–191. [CrossRef]
48. Dalgleish, G.S. The Sanayta lock and bride-regulator (Including plate at back of volume). In *Minutes of the Proceedings of the Institution of Civil Engineers*; ICE Publishing: Washington, DC, USA, 1901.

heritage

MDPI

Article

Comparative Inhibition Study by Nanomaterial, Plant Extract and Chemical Microcide on the Screaming Mummy in Egyptian Museum Store

Sahar Ismael [1,*], Ali Omar [2] and Manal Maher [3]

1 Conservation Department, Faculty of Archaeology, Fayoum University, Fayoum 63514, Egypt
2 Microbiology Department, Conservation Center, Grand Egyptian Museum, Giza 12572, Egypt; Dr.Ali_omar@yahoo.com
3 Nanotechnology Center, Cairo University, Giza 12585, Egypt; manal.a.maher@gmail.com
* Correspondence: sma17@fayoum.edu.eg

Abstract: Mummies in museums are exposed to different deterioration factors like microorganisms, especially unwrapped mummies, such as the screaming mummy. This screaming mummy in the store of the Egyptian museum is suffering from stains due to microbial infection. There are three trends of materials to inhibit microbial growth: nano materials, plant extraction and chemical materials. This research compares three materials representing the three trends such as nano zinc oxide (ZnO-NPs), *Ceratophyllum demersum* and 4-chloro-m-cresol, respectively. Microorganisms, isolated from the degraded mummy, were identified with an optical microscope and ribosomal ribonucleic acid (rRNA) analysis to guarantee identification accuracy. Results indicated that the bacteria in the mummy are *Bacillus jeotgali, Kocuria turfanensis, Microbacterium imperial, Micrococcus luteus* and *Bacillus megaterium.* Fungi are *Monascus pallens* and *Rhizopus oryzae.* The results of minimum inhibitory concentration (MIC) illustrated that the best concentrations for the bio treatment of isolated microorganisms is plant extract (*Ceratophyllum demersum*) at 600 ppm/100 mL, followed by 4-chloro-m-cresol at 600 ppm/100 mL and finally nano zinc oxide at 700 ppm/100 mL.

Keywords: nano zinc oxide; *Ceratophyllum demersum*; 4-chloro-m-cresol; bacteria; fungi

Citation: Ismael, S.; Omar, A.; Maher, M. Comparative Inhibition Study by Nanomaterial, Plant Extract and Chemical Microcide on the Screaming Mummy in Egyptian Museum Store. *Heritage* **2021**, *4*, 2481–2493. https://doi.org/10.3390/heritage4030140

Academic Editors: Christofilis Maggidis and Omar Abdel-Kareem

Received: 18 August 2021
Accepted: 13 September 2021
Published: 16 September 2021

1. Introduction

Mummification is a process that was done to preserve dead body in ancient Egypt by embalming and wrapped with linen bandages [1]. The ancient Egyptians did this process due to religious concepts, such as the afterlife [2]. Mummification includes the treatment of the body with oils, resin and wax wrapped with linen and placing it in a sealed coffin to ensure the preservation of the body [3]. Screaming (Glaring) mummies do not have good embalming. There are two types of screaming mummies: the first and most famous ones were murdered because of committing a crime, so they do not have good mummification or linen wrapping. The second type died as a result of a disease and all the usual embalming steps are done for them [4].

The mummy suffers from several factors that cause deterioration in museums, for example microorganisms [5], especially fungi more than bacteria. These organisms produce enzymes that cause staining and a decayed mummy [6]. The damage percentage by microorganisms depends on species, the surrounding climate (humidity, temperature) and the quality of the embalming [7,8]. Screaming mummies of the first type are more exposed to deterioration by fungi and bacteria because of bad mummification especially because they were not wrapped in linen strips.

The main concern of the study is the mummy of an unknown woman (Figure 1) from the store of the Egyptian Museum in Cairo, Egypt. The mummy's number is CIT 8; it was discovered in an unknown tomb by Winlock's excavations at ThT71. Winlock's written note cards say: S.A.K. (1935–1936) Servants Cemetery burial No. 2 (SENMUT) of the 18th Dynasty. The height of the mummy is 158 cm (height from the heel to the crown of the head), and the age was estimated in the elderly.

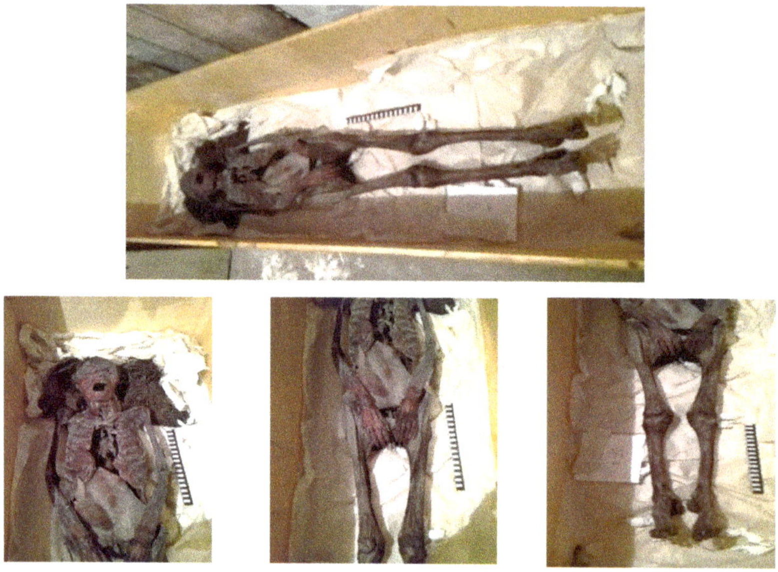

Figure 1. The screaming mummy from the 18th dynasty in the Egyptian museum store.

Lately, nano materials are used as an antimicrobial agent [9]. Nanomaterials (1–100 nm) have a lot of properties: reducing UV effects, consolidation, cleaning, deacidification and inhibition of microbial growth [10–13]. Nano zinc oxide (ZnO-NPs) is known as a good antimicrobial Aaent, [14] especially for leather [15]—thus, it was chosen for the current study.

Ancient Egyptians used plants that have antimicrobial activities (due to contents of phenols and esters) and essential oils for mummification [16]. Currently, plant extracts and essential oils are used as antimicrobial agents in different ways for archaeological objects. A plant extract is prepared byusing different parts of the plant (often the leaves) soaked in 100 mL of water and 100 mL of ethanol, then evaporated to obtain a substance dissolved in 200 mL of sterile distilled water. Essential oils are more effective for inhibiting microbial growth, but they are not suitable for every archaeological materials. They can cause darkening in color [17]. Scientific studies have mentioned that camphor and thyme are more effective than other essential oils [18,19]. Another study proved that plant extract and nano materials did not change or produce a reaction with archaeological objects' color [20]. Thus, it was chosen for the current study.

Some of hydrophytes in the Nile River have antimicrobial activity, including *Ceratophyllum demersum* [21]. *Ceratophyllum demersum* is a submerged aquatic plant from Saluga and Ghazal island [22]; they are protected areas in Aswan, Egypt [23].

There are a lot of chemical materials used as antimicrobials and pesticides. Some of them could be used in the treatment of archaeological objects. It is preferable to use safe antimicrobials for objects as a restorer, such as 4-chloro-m-cresol [24]. 4-chloro-m-cresol is considered a safe material to be used even in cosmetic products (up to 0.2%). This material is a synonym for p-chloro-m-cresol [25,26].

The current research aims to study microorganisms isolated from the screaming mummy and to study its biological activities. The study also aims to compare the inhibition of three materials: nano zinc oxide, *Ceratophyllum demersum* and 4-chloro-m-cresol representing, respectively. These three materials are for inhibiting microbial attacks, and are applied to the microorganisms isolated from the screaming mummy.

2. Materials and Methods

2.1. Materials

The nano material used in this study, zinc oxide (ZnO-NPs) < 50 nm particle size (BET), >97%, was purchased from Sigma-Aldrich Company, No: 677450. The plant extract was *Ceratophyllum demersum*. The plant is a dominant species within the Nile River system. The plant was collected from sites located around Isis island (24°04.646′ N; 32°52.701′ E) and Saluga and Ghazal island (24°04.328′ N; 32°52.279′ E). It was prepared by washing the plant, then it was dried and ground into a powder. The powder was then left in methanol. The chemical antimicrobial was 4-chloro-m-cresol, purchased from Loba Chemie Pvt. Ltd., Mumbai, India, No: 02750. The media used for isolation and purification were cellulose, protein and nutrient agar.

2.2. Methods

A collection of swabs was taken from the screaming mummy in the Egyptian Museum store, specifically from skin areas including the head, wig, rib cage, abdominal and pelvic cavity, right leg, left leg and mouth of the mummy as the presence of microbial lesions is clear in the form of brown spots (Figure 2). Microbial swabs were cultured on plates of cellulose, protein and nutrient agar media. Regarding isolation and purification, the growths that presented in the Petri dishes were taken separately and cultured on the same previous media in order to obtain the microorganisms in pure forms to complete the steps of characterization.

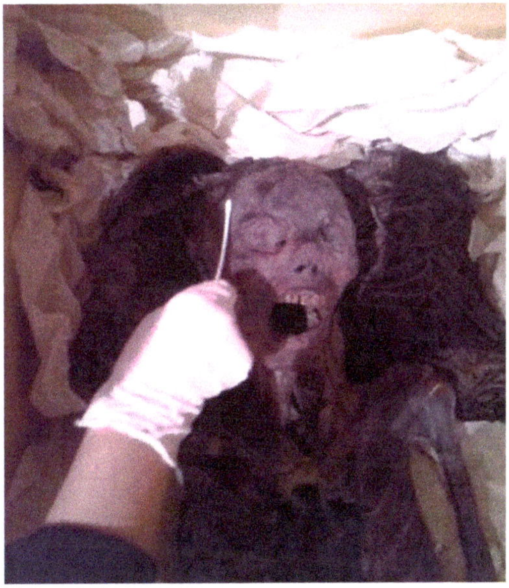

Figure 2. Collecting swabs from the mummy.

Microbial colonies that grew on the incubated plates were sub-cultured into separate fresh sterile cellulose agar and nutrient agar plates and then incubated to obtain pure cultures of causative microorganisms. The purified isolates were kept in slants and stored for characterization. Identification of all microbial isolates was carried out at the Laboratory of Microbiology, Grand Egyptian Museum, Ministry of Antiquities with optical microscopy and with sequencing of rRNA genes at Solgent Company, South Korea.

Bacterial identification was performed with a molecular approach. Bacterial isolates were cultured in sterile test tubes containing 10 mL of nutrient broth medium [27]. Cultures were incubated at 28 °C for 48 h. A small amount of bacterial culture was scraped by sterile spatula suspended in 100 µL of sterile distilled water in 2 mL of sterile vials and boiled at 100 °C for 15 min. Bacterial DNA was extracted and isolated using SolGent purification bead. Prior to sequencing, the ribosomal rRNA gene was amplified using the polymerase chain reaction (PCR) technique in which two universal bacterial primers 27F (forward) and 1492R (reverse) and were incorporated in the reaction mixture. Primers used for gene amplification have the following composition: 27F(5′AGAGTTTGATCMTGGCTCAG) and 1492R(5′TACGGYTACCTTGTTACGACTT).

The purified PCR products (amplicons) were reconfirmed using a size nucleotide marker (100 base pairs) by electrophoreses on 1% agarose gel. Then, these bands were eluted and sequenced with the incorporation of dideoxynucleotides (dd NTPs) in the reaction mixture. Each sample was sequenced in the sense and antisense directions using the same primers [28]. Sequences was further analyzed using Basic Local Alignment Search Tool (BLAST) from the National Center of Biotechnology Information (NCBI) website. Phylogenetic analysis of sequences was done with the help of Meg Align (DNA Star) software version 5.05. Identification of bacterial isolates was done by sequencing of rRNA genes at the Solgent Company, South Korea.

Fungal isolates can be identified with different methods (morphological, molecular approach and biochemical), but the molecular approach was used because of its high accuracy compared to the other methods. Fungal isolates were cultivated in Petri plates containing 20 mL of Czapek's yeast extract agar with 20% sucrose (CY20S) medium at 28 °C for 7 days [29]. A small amount of fungal culture was scraped by sterile spatula suspended in 100 µL sterile distilled water in 2ml sterile vials and boiled at 100 °C for 15 min. Fungal DNA was extracted and isolated using SolGent purification bead. Prior to sequencing, the ribosomal rRNA gene was amplified using the polymerase chain reaction (PCR) technique in which two universal fungal primers were sequenced with ITS1 and ITS4 primers [30], which were incorporated in the reaction mixture. Primers used for gene amplification have the following composition: ITS1 (5′-TCCGTAGGTGAACCTGCGG3′) and ITS4 (5′-TCC TCCGCTTATTGATATGC-3′) for fungi.

The purified PCR products (amplicons) were reconfirmed using a size nucleotide marker (100 base pairs) by electrophoreses on 1% agarose gel. The amplicons were sequenced with the incorporation of dideoxynucleotides (dd NTPs) in the reaction mixture. Sequences were further analyzed using Basic Local Alignment Search Tool (BLAST) from the National Center of Biotechnology Information (NCBI) website. Phylogenetic analysis of sequences was done with the help of MegAlign (DNA Star) software version 5.05. Identification of fungal isolates was done by sequencing of rRNA gene at the Solgent Company, South Korea.

In order to determine cellulose (s) and protease activity of the isolated microorganisms we used the following steps:

- Enzyme Production

Production was carried out in 250 mL conical flasks. Each contained 100 mL of the production medium for producing cellulase and protease enzymes and the main source of carbon (sucrose) was replaced with 10g of cellulose and 10 g of gelatin, respectively. Flasks were sterilized at 121 °C for 15 min. After that, cooling inoculated with 2 ml of standard inoculums of each isolate. The inoculated flasks were incubated at 28–30 °C for the proper time. At the end of the incubation period, the liquid cultures were centrifuged at 3000 RPM for 15 min. The supernatant was taken for determination of enzyme (cellulase and protease) activity as described below.

- Enzymes Assay

The cup plate clearing zone technique (CCZ) was used for assaying activities of cellulase and protease enzymes. The procedure was carried out by pouring 20 mL aliquots of the detection medium [31] into a sterile Petri dish and it was then allowed to solidify. A sterile cork borer (15 mm diameter) was used to make three cups in each plate and 0.1 mL of the supernatant (cell free enzyme) of each isolate was placed into the three cups. Plates were incubated at 30 °C for 24 h after which the plates were flooded with a Lugol's iodine solution to assay cellulose and with an acid mercuric chloride solution for protease assay. Enzyme activities were compared based on the diameter (mm) of clear zone.

2.2.1. Determine Minimal Inhibitory Concentration (MIC) of Antimicrobial Agents against the Isolated Microorganisms

The three substances categories microcides chosen were: nano material, plant extract and chemical antimicrobial (ZnO NPs, *Ceratophyllum demersum* and 4-chloro-m-crysol) prepared to determine their minimal inhibitory concentrations. A stock solution of each microcide was prepared by dissolving 1 g/L ethyl alcohol (95%) in the case of 4-chloro-m-crysol and nano zinc oxide and 1 mL/L of ethyl alcohol (95%) in the case of *Ceratophyllum demersum*. Gradient concentrations of each microcide (400, 500, 600 and 700 ppm) were prepared by diluting the stock solution with alcohol.

One milliliter of spore suspension was spread on nutrient agar and Dox's agar plate. The controls were the solvents used for every extract. Plates were left to be dry. A cork borer was used to make three pores in each plate and 100 µL of the three concentrations was added for each microcide materials. Concentrations start with 400 ppm/100 mL. The zones of growth inhibition around the wells were measured after 18 to 24 h of incubation at 37 °C for bacteria and 48 to 96 h for fungi at 28 °C. The sensitivities of the microorganism species to the test substances were determined by the sizes of inhibition zones (including the diameter of well) on the agar surface around the wells. Values less than 15 mm were considered as not active against microorganisms compared to control plates that used ethyl alcohol (dissolvent solution of microcides) alone.

2.2.2. Evaluate Treatment for Isolated Microorganisms from Mummy

Microbial growth was examined by taking of swabs from each treated plates after 48 h, 3 months and 6 months. The swabs were cultured in Dox's medium for fungi and Nutrient agar for bacteria. A concentration of the best microcide will be used.

3. Results

3.1. Identification of the Microbial Isolates

3.1.1. Optical Microscope Identification

The resulting microbial colonies were subjected to preliminary characterization depending on the type of organism, as mentioned previously. The following genera were identified: *bacillus, coccus* and *Rhizopus* (Table 1 and Figures 3 and 4).

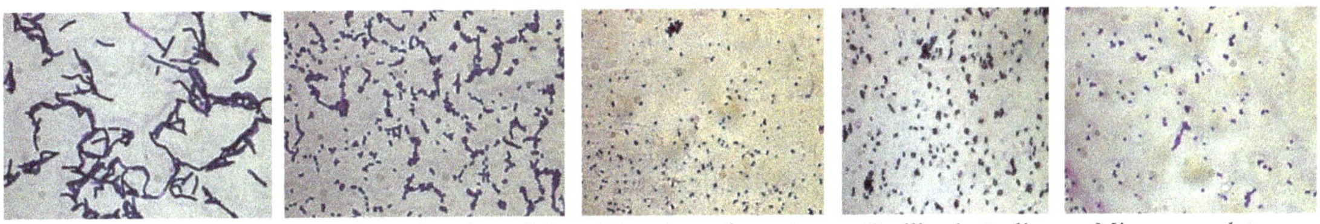

Bacillus megaterium *Microbacterium imperiale* *Kocuria turfanensis* *Bacillus jeotgali* *Micrococcus luteus*

Figure 3. Bacteria isolated from mummy with optical microscope at 1000× magnification.

Based on the results, it can be seen that the genus *bacillus* was the dominant genus in seven sites having 33.3% of the total bacterial isolates and the genus *Rhizopus* was the dominant genus in seven sites having 55.2% of the total fungal isolates.

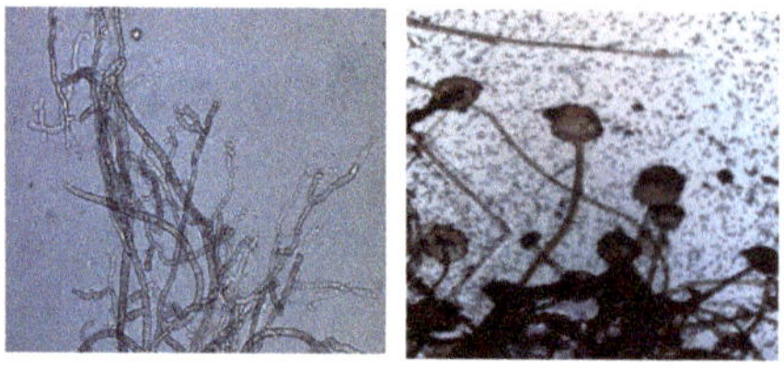

Monascus pallens *Rhizopus oryzae*

Figure 4. Fungi isolated of mummy with optical microscope at 400× magnification.

Table 1. Isolated microorganisms grow on different media (cellulose, protein and nutrient agar).

Swab Cite	Media Used		
	Cellulose Agar	**Protein Agar**	**Nutrient Agar**
Head	Fungus 1 *Rhizopus oryzae*	G+ve *Shortbacilli* spore former1	G+ve *Shortbacilli* spore former2 *Microbacillus* spp.
Wig	*Microbacillus* spp. Fungus 1	*Microbacillus* spp. G+ve *Shortbacilli* spore former3	G+ve *Shortbacilli* spore former3
Rib cage	*Rhizopus oryzae*	G+ve *Shortbacilli* spore former2	G+ve *Shortbacilli* spore former4 *Microbacillus* spp.
Abdominal and pelvic cavity	Fungus 1	*Microbacillus* spp.	G+ve *Shortbacilli* spore former3 *Micrococus* spp.
Right leg	*Rhizopus oryzae*	*Microbacillus* spp. G+ve *Shortbacilli* spore former1	G+ve *Shortbacilli* spore former3
Left leg	*Rhizopus oryzae* *Microbacillus* spp.	G+ve *Shortbacilli* spore former3 G+ve *Shortbacilli* spore former4	G+ve *Shortbacilli* spore former3 *Microbacillus* spp.
Mouth	*Rhizopus oryzae* Fungus 1	*Micrococus* spp.	G+ve *Shortbacilli* spore former2 *Micrococus* spp.

3.1.2. rRNA Sequencing Identification

Based on the results in Tables 2 and 3, it was found that the bacterial isolates were identified according to thew molecular approach by 16S rRNA sequencing analysis compared with closely related strains accessed from the Gen Bank, such as *Bacillus jeotgali, Kocuria turfanensis, Microbacterium imperial, Micrococcus luteu* and *Bacillus megaterium*. Fungal isolates were identified according to the molecular approach by ITS region of rDNA sequencing analysis and compared with closely related strains accessed from the Gen Bank, such as *Monascus pallens* and *Rhizopus oryzae*.

3.2. *Determination of Cellulases and Protease Produced by the Isolated Microorganisms by Cup Plate Technique*

Data recorded in Table 4 and illustrated by Figure 5 indicate that the tested microorganisms proved the various abilities to produce the cellulases and protease enzymes. Microorganisms are different in the degree of decomposition of proteins and cellulose. The tabulated data show the highest activity of cellulolytic activity observed in *Bacillus jeotgali* and a highest activity of proteolytic activity observed in *Microbacterium imperial*.

Table 2. Identification of bacterial isolates obtained from deteriorated mummy by 16S rRNA sequencing analysis compared with closely related strains accessed from the Gen Bank.

Swab Cite	Identification	Closely Related Microbial Strains Accessed from Gen Bank		
		Strain No.	Accession No.	Similarity
Head	*Bacillus jeotgali* *Kocuria turfanensis* *Microbacterium imperiale*	YKJ-10 HO-9042 DSM 20530	NR025060T NR043899T NR026161T	99.93% 88.22% 99.86%
Wig	*Microbacterium imperiale* *Bacillus jeotgali*	DSM 20530 YKJ-10	NR026161T NR025060T	99.86% 99.93%
Rib cage	*Microbacterium imperiale* *Bacillus jeotgali*	DSM 20530 YKJ-10	NR026161T NR025060T	99.86% 99.93%
Abdominal and pelvic cavity	*Microbacterium imperial* *Micrococcus luteu* *Bacillus jeotgali*	DSM 20530 MB5 YKJ-10	NR026161T MH450098 NR025060T	99.86% 99.71% 99.93%
Right leg	*Bacillus megaterium* *Microbacterium imperial* *Kocuria turfanensis*	NBRC 15308 = ATCC 14581 DSM 20530 HO-9042	NR112636T NR026161T NR043899T	100% 99.86% 88.22%
Left leg	*Microbacterium imperial* *Bacillus jeotgali* *Bacillus megaterium*	DSM 20530 YKJ-10 NBRC 15308 = ATCC 14581	NR026161T NR025060T NR112636T	99.86% 99.93% 100%
Mouth	*Bacillus jeotgali* *Micrococcus luteus*	YKJ-10 MB5	NR025060T MH450098	99.93% 99.71%

Table 3. Identification fungal isolates obtained from deteriorated mummy ITS region of rDNA sequencing analysis compared with closely related strains accessed from the Gen Bank.

Scheme	Identification	Closely Related Microbial Strains Accessed from Gen Bank		
		Strain No.	Accession No.	Similarity
Head	*Monascus pallens* *Rhizopus oryzae*	IMI 356820 IDI 354840	GU733328 GU732332	100% 99.88%
Wig	*Microbacterium imperial* *Monascus pallens*	DSM 20530 IMI 356820	NR026161T GU733328	99.86% 100%
Rib cage	*Rhizopus oryzae*	IDI 354840	2GU732332	99.88%
Abdominal and pelvic cavity	*Monascus pallens*	IMI 356820	GU733328	100%
Right leg	*Rhizopus oryzae*	IDI 354840	GU732332	99.88%
Left leg	*Microbacterium imperial* *Rhizopus oryzae*	DSM 20530 IDI 354840	NR026161T GU732332	99.86% 99.88%
Mouth	*Monascus pallens* *Rhizopus oryzae*	IMI 356820 IDI 354840	GU733328 GU732332	100% 99.88%

3.3. The MIC of Isolated Micro-Organisms

In this part of the study, the three microcides were applied to the isolated microorganisms to determine the MIC. Nano zinc oxide showed no response at various concentrations (400, 500 and 600 ppm/100 mL) in all isolates, except for at the concentration 700 ppm/100 mL that showed a response to all isolates. Thus, it could be reported that using nano zinc oxide at concentrations of 700 ppm/100 mL gave the diameter of a clear zone that ranged from 18–25 mm.

Table 4. Determine of cellulases and protease produced by the isolated microorganisms using cup plate technique.

Microbial Isolates	Diameter of Clearing Zone (mm)	
	Cellulases	Protease
Monascus pallens	23	22
Rhizopus oryzae	24	21
Bacillus jeotgali	28	27
Kocuria turfanensis	19	20
Microbacterium imperial	26	32
Micrococcus luteus	24	25
Bacillus megaterium	22	21

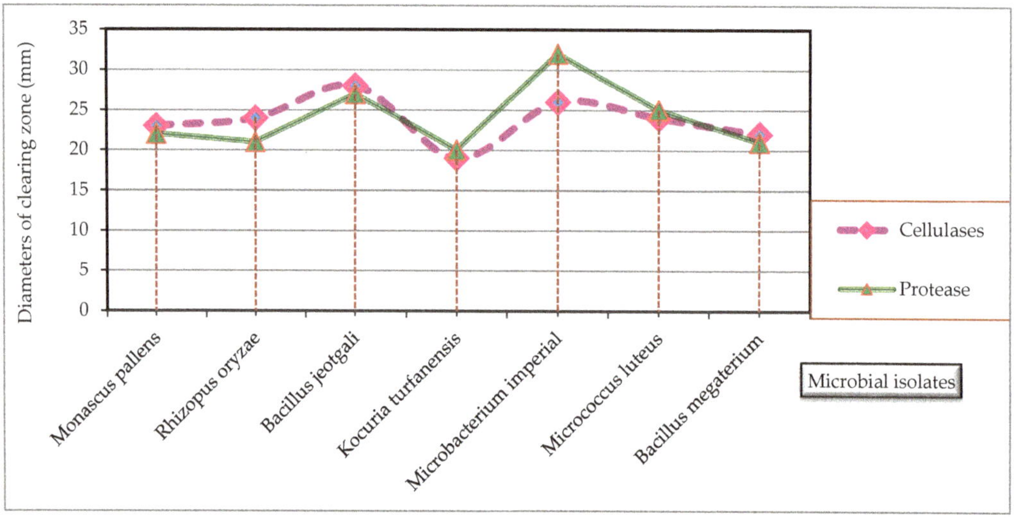

Figure 5. Cellulase(s) and protease produced by isolated microorganisms using cup plate technique.

Natural substance plant extract (*Ceratophyllum demersum*) had a good response at 600 ppm/100 mL concentration, which gave a response in all isolates. Therefore, it could be reported that using a natural substance at 600 ppm/100 mL gave the diameter of a clear zone ranging from 25–48 mm.

In the case of 4-chloro-m-cresol, no response could be detected at 400 ppm/100 mL concentration of all isolated microorganisms. It showed a response at 500 ppm/100 mL with most isolated microorganisms, but the good response was at 600 ppm/100 mL, where the diameter of a clear zone ranged from 24–45 mm.

These data can be used to recommend the best concentrations of a specific microbicide for the microbial treatment of an infected mummy. For instance, a natural substance plant extract (*Ceratophyllum demersum*) at 600 ppm/100 mL is sufficient to inhibit all isolated microorganisms followed by 4-chloro-m-cresol at 600 ppm/100 mL and Nano zinc oxide at 700 ppm/100 mL. These results are listed in Table 5 and shown in Figures 6 and 7.

3.4. Evaluation of Different Methods for Treatment of Deteriorated Mummy

The plant extract was the best to be applied using the fumigation method (30 mL/m^3). A concentration of 600 ppm/100 mL of plant extract was used to inhibit microbial growth on the mummy. Microbial growth was examined by taking of swabs from each treated specimen after 48 h, 3 months and 6 months. The swabs were cultured in Dox's agar medium for fungi and Nutrient agar for bacteria (Table 6).

Table 5. Determine inhibition zone (mm) of microbial species grown on Czapek agar and Nutrient agar as affected by three microcides: nano zinc oxide, natural plant extract (*Ceratophyllum demersum*) and 4-chloro-m-cresol.

Microorganisms	Inhibition Zone by mm at Different Concentrations (ppm/100 mL).									
	Nano Zinc Oxide				Plant Extract			4-chloro-m-cresol		
	400	500	600	700	400	500	600	400	500	600
Monascus pallens	0	0	0	18	0	20	37	0	25	35
Rhizopus oryzae	0	0	0	25	0	30	48	0	30	45
Bacillus jeotgali	0	0	0	18	0	0	28	0	0	27
Kocuria turfanensis	0	0	0	22	0	27	40	0	25	28
Microbacterium imperial	0	0	0	21	0	19	25	0	20	24
Micrococcus luteus	0	0	0	22	0	33	43	0	25	32
Bacillus megaterium	0	0	0	21	0	25	36	0	25	33

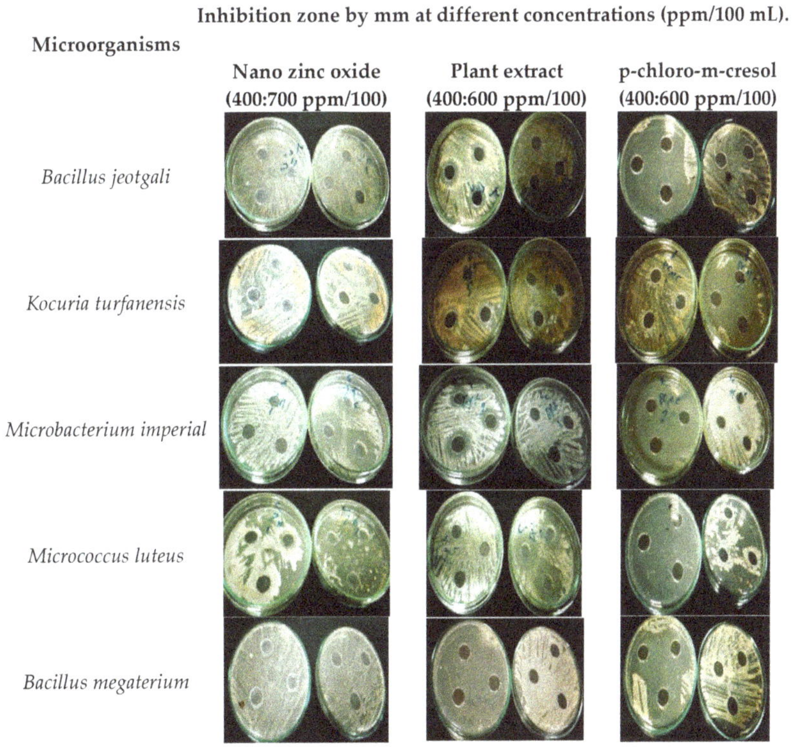

Figure 6. Effect of the three microcides: nano zinc oxide, natural plant extract (*Ceratophyllum demersum*) and 4-chloro-m-cresol on bacteria isolated from the mummy.

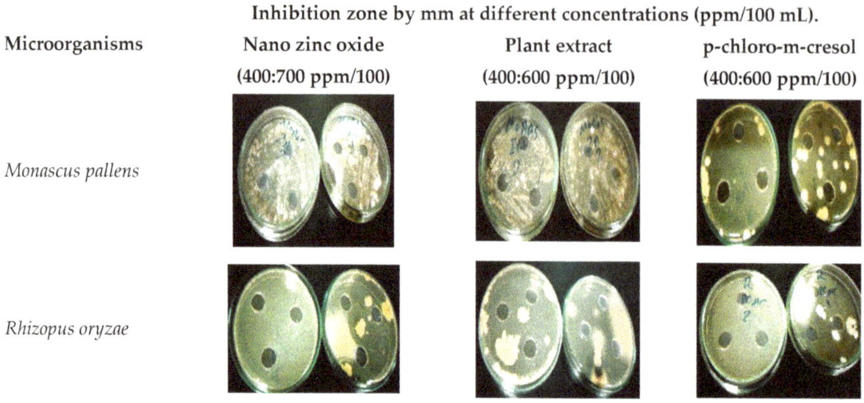

Figure 7. Effect of the three microcides: nano zinc oxide, natural plant extract (*Ceratophyllum demersum*) and 4-chloro-m-cresol on fungi isolated from the mummy.

Table 6. The effect of applying *Ceratophyllum demersum* at (600 ppm/100 mL) on microbial growth isolated from mummy.

Growth Detection Time	Growth of the Infectious Isolates after Treatment with (*Ceratophyllum demersum* L.) at 600 ppm/100 mL
48 h	No Growth
3 Month	No Growth
6 Month	No Growth

4. Discussion

Results show that the concentration 500 ppm/100 mL of nano zinc oxide does not have any effect, but the plant extract had the highest inhibition zone with most isolated microorganisms. In the concentration of 600 ppm/100 mL, there is no response for nano zinc oxide as well, although the plant extract and chemical microcide have a good inhibition zone. Therefore, nano zinc oxide in the concentration 700 ppm/100 mL was compared to a concentration of 600 ppm/100 mL of the for natural Plant extract (*Ceratophyllum demersum*) and chemical microcide, 4-chloro-m-cresol (Figures 8 and 9).

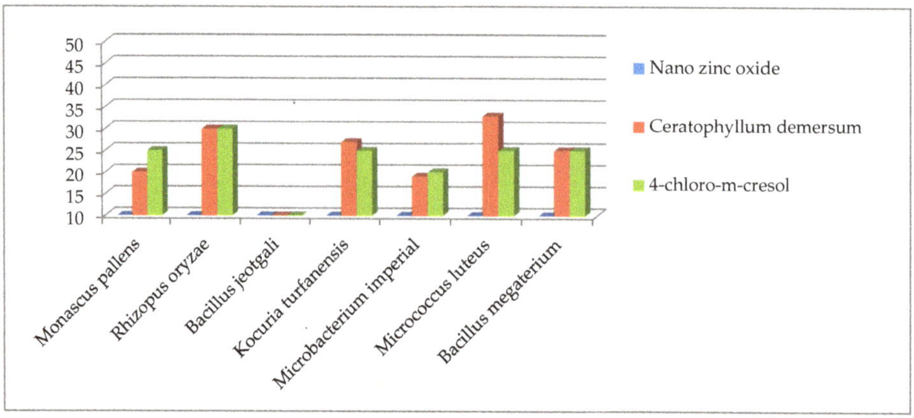

Figure 8. Comparison of the inhibition zone (mm) of microbial species grown on Czapek agar and Nutrient agar as affected by the 500 ppm/100 mL concentration of three microcides: nano zinc oxide, *Ceratophyllum demersum* and 4-chloro-m-cresol.

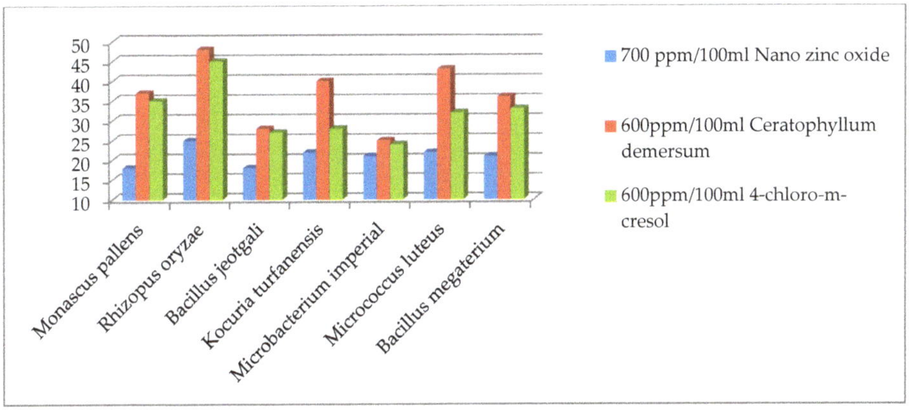

Figure 9. Comparison of the inhibition zone (mm) of microbial species grown on Czapek agar and Nutrient agar as affected by the 700 ppm/100 mL nano zinc oxide and 600 ppm/100 mL concentration of *Ceratophyllum demersum* and 4-chloro-m-cresol.

The chemical microcide (4-chloro-m-cresol), applied by brush or spray, could be concentrated in places and forgotten in others. This material could have an effect on the restorer's health if they do not wear a mask. For that reason, the chemical microcide (4-chloro-m-cresol) was not used. The nano material (ZnO NPs) is safer than 4-chloro-m-cresol for the restorer, but could be also concentrated in some places and cause a whitening of the mummy's appearance, especially if the mummy does not have wrapped linen.

Natural plant extracts (*Ceratophyllum demersum*) are used by fumigation in closed areas. A tent from polyethylene (plastic) was made and the mummy was put inside to be treated. Then, determining the volume of the tent in cubic meters and the corresponding volume of the plant extract is put into a Petri dish at the ratio of 30 mL/m^3. After that, the tent was sealed and left for 48 h. Microbial growth was examined by taking swabs from each treated specimen after 48 h, 3 months and 6 months. The swabs were cultured in Dox's medium for fungi and Nutrient agar for bacteria. This fumigation method is the best for archaeological objects that have many details, such as the screaming mummy. This mummy does not have wrapped linen strips. In addition, the *Ceratophyllum demersum* does not cause any change to the archaeological material's color and does not harm the restorer's health. This result is similar to the treatment of rare books [32].

5. Conclusions

The ancient mummies preserved in the Egyptian Museum store deteriorated due to the poor level of preservation conditions. Microorganisms isolated from the screaming mummy were identified based on ribosomal RNA analysis. Results show the presence of bacteria: *Bacillus jeotgali, Kocuria turfanensis, Microbacterium imperial, Micrococcus luteus* and *Bacillus megaterium*. The fungi are *Monascus pallens* and *Rhizopus oryzae*.

The isolated microorganisms produced activity in cellulose and protease enzymes that can cause biodeterioration of ancient mummies. The comparison between the three types of microcides material showed that the plant extract, *Ceratophyllum demersum*, at a concentration of 600 ppm/100 mL is the best. It is sufficient to inhibit all isolated microorganisms.

Author Contributions: Conceptualization, S.I.; methodology, S.I. and A.O.; software, S.I.; investigation, S.I.; data curation, S.I., A.O. and M.M.; writing—original draft preparation, S.I. and A.O., writing—review and editing, S.I.; visualization, S.I. All authors have read and agreed to the published version of the manuscript.

Funding: This research received no external funding.

Institutional Review Board Statement: Not applicable.

Informed Consent Statement: Not applicable.

Data Availability Statement: Not applicable.

Conflicts of Interest: The authors declare no conflict of interest.

References

1. Museum of Natural History Smithsonian Institution. Egyptian Mummies. 2012. Available online: https://www.si.edu/spotlight/ancient-egypt/mummies (accessed on 15 May 2021).
2. Abdel-Maksoud, G.; El-Amin, A. A review on the materials used during the mummification processes in ancient Egypt. *Mediterr. Archaeol. Archaeom.* **2011**, *11*, 129–150.
3. Magdy, M.; Ismail, M.; Issa, Y.; Abdel-Maksoudc, G.; Ibrahim, M. An analytical study for understanding the degradation process of a late period mummy. *Adv. Res. Conserv. Sci.* **2020**, *1*, 13–30.
4. Hawass, Z.; Saleem, S. Computed tomography examination of the screaming mummy "Unknown-Woman-A". *Egypt. J. Radiol. Nucl. Med.* **2020**, *51*, 139. [CrossRef]
5. Arya, A.; Shah, A.R.; Sadasivan, S. Indoor aeromycoflora of Baroda museum and deterioration of Egyptian mummy. *Curr. Sci.* **2001**, *81*, 793–799.
6. Naji, M.K.; Abdullah, Q.Y.; AL-Zaqri, A.Q. Evaluating the biodeterioration enzymatic activities of fungal contamination isolated from some ancient Yemeni mummies preserved in the national museum. *Biochem. Res. Int.* **2014**, *2014*, 481508. [CrossRef]
7. Omar, A.M.; Taha, A.S.; Mohamed, A.A. Microbial deterioration of some archaeological artifacts: Manipulation and treatment. *Eur. J. Exp. Biol.* **2018**, *8*, 21. [CrossRef]
8. Elamin, A.; Takatori, K.; Matsuda, Y.; Tsukada, M.; Kirino, F. Microbialogical, morphological and spectroscopic study on the effect of resinous materials in the preservation of wrapping textiles of mummies. *Mediterr. Archaeol. Archaeom.* **2018**, *18*, 1–10. [CrossRef]
9. David, M.E.; Ion, R.M.; Grigorescu, R.M.; Iancu, L.; Andrei, E.R. Nanomaterials used in conservation and restoration of cultural heritage: An up-to-date overview. *Mater. J.* **2020**, *13*, 2064. [CrossRef]
10. Bonini, M.; Lenz, S.; Giorgi, R.; Baglioni, P. Nanomagnetic sponges for the cleaning of works of art. *Langmuir J.* **2007**, *23*, 8681–8685. [CrossRef]
11. Giorgi, R.; Chelazzi, D.; Fratini, E.; Langer, S.; Niklasson, A.; Rådemar, M.; Svensson, J.E.; Baglioni, P. Nanoparticles of calcium hydroxide for wood deacidification: Decreasing the emissions of organic acid vapors in church organ environments. *J. Cult. Herit.* **2009**, *10*, 206–213. [CrossRef]
12. Helmi, F.M.; Ali, N.M.; Ismael, S.M. Nanomaterials for the inhibition of microbial growth on Ancient Egyptian Funeral masks. *Mediterr. Archaeol. Archaeom.* **2015**, *15*, 87–95. [CrossRef]
13. Sricharussin, W.; Threepopnatkul, P.; Neamjan, N. Effect of various shapes of zinc oxide on cotton fabric for UV-Blocking and anti-bacterial properties. *Fibers Polym. J.* **2011**, *12*, 1037–1041. [CrossRef]
14. Taghizadeh, S.M.; Lal, N.; Ebrahiminezhad, A.; Moeini, F.; Seifan, M.; Ghasemi, Y.; Berenjian, A. Green and economic fabrication of Zinc Oxide (ZnO) nanorods as a broadband UV blocker and antimicrobial agent. *Nanomater. J.* **2020**, *10*, 530. [CrossRef]
15. Khamseh, H. Conservation of leather historical object by Nano technology in archaeology found (Sample study cover leather book). *J. Fundam. Appl. Sci.* **2016**, *8*, 276–283. [CrossRef]
16. Mansour, M.M.; EL-Hefny, M.; Salem, M.; Ali, H.M. The Biofungicide activity of some plant essential oils for the cleaner production of model linen fibers similar to those used in ancient Egyptian mummification. *Process. J.* **2020**, *8*, 79. [CrossRef]
17. Othman, M.; Saadat, H.; Matsuda, Y. Antifungal activity of some plant extracts and essential oils against Fungi-Infested organic archaeological artefacts. *Archaeometry* **2020**, *62*, 187–199. [CrossRef]
18. Elsayed, Y.; Shabana, Y. The effect of some essential oils on aspergillus Niger and Alternaria Alternata infestation in archaeological oil paintings. *Mediterr. Archaeol. Archaeom.* **2018**, *18*, 71–87. [CrossRef]
19. Geweely, N.S.; Afifi, H.A.; Ibrahim, D.M.; Soliman, M.M. Efficacy of essential oils on Fungi isolated from archaeological objects in Saqqara excavation, Egypt. *Geomicrobiol. J.* **2019**, *36*, 148–168. [CrossRef]
20. Khalaphallah, R.; El-Derby, A. The effect of nano TiO_2 and plant extracts on microbial strains isolated from Theban ancient Egyptian royal tomb painting. *Afr. J. Microbiol. Res.* **2015**, *9*, 1424–1430. [CrossRef]
21. Metwally, F.E.; Mohamed, A.A.; Mahalel, U.A.; Sheded, M.G. Evaluation of certain cosmopolitan hydrophytes in the Nile River, Aswan district for their ecological and bioactivity potentials: A review. *Int. J. Sci. Technol. Res.* **2020**, *9*, 1595–1606, ISSN 2277–8616.
22. Omar, A.M.; Salah, T.A.; Mohamed, A.A.; Sheded, M.G. Attenuation of microbial induced deterioration of cellulose fibers by hornwort (*Ceratophyllum demersum* L.) methanolic extract. *Int. J. Biol. Res.* **2017**, *5*, 48–58. [CrossRef]
23. Abdel-wahab, H.M.; Abdel-kader, A.E.; Yousef, R.S. Population dynamics of terrestrial insects in Saluga and Ghazal protected area, Aswan, Egypt. *Egypt. Soc. Environ. Sci.* **2019**, *18*, 117–123.
24. Omar, A.; Taha, A.; El-Wekeel, F. Microbial degradation of ancient textiles in the Egyptian textile museum and methods of its control. *Egypt. J. Archaeol. Restor. Stud.* **2019**, *9*, 27–37.

25. US EPA (United States Environmental Protection Agency). Reregistration Eligibility Decision (RED) p-Chloro-m-cresol, Prevention, Pesticides and Toxic Substances. EPA-738-R-96-008. 1997. Available online: https://archive.epa.gov/pesticides/reregistration/web/pdf/3046red.pdf (accessed on 15 May 2021).
26. Andersen, F.A. Final report on the safety assessment of p-chloro-m-cresol. *Int. J. Toxicol.* **1997**, *16*, 268.
27. Zimbro, M.J.; Power, D.A.; Miller, S.M.; Wilson, G.E. *and Johnson, J.A. Difco & BBL Manual of Microbiological Culture Media*, 2nd ed.; Becton, Dickinson and Company: New Jersey, MD, USA, 2009.
28. Lane, D.J. ; *16S/23S rRNA Sequencing. Nucleic Acid Techniques in Bacterial Systematics*; John Wiley and Sons: Chichester, UK, 1991; pp. 115–175. Available online: http://lycofs01.lycoming.edu/~{}newman/NovelMicrobe/16SrRNAprimers.pdf (accessed on 15 June 2021).
29. Pitt, J.I.; Hocking, A.D. *Fungi and Food Spoilage*, 3rd ed.; Springer Science & Business Media LLC: Berlin/Heidelberg, Germany, 2009.
30. White, T.J.; Bruns, T.; Lee, S. and Taylor, J. Amplification and direct sequencing of fungal ribosomal RNA genes for phylogenetics. In *PCR Protocols: A Guide to Methods and Applications*; Innis, M.A., Gelfand, D.H., Sninsky, J.J., White, T.J., Eds.; Academic Press: San Diego, CA, USA, 1990; pp. 315–322.
31. Pangallo, D.; Simonovičová, A.; Chovanova, K.; Ferianc, P. Wooden art objects and the museum environment: Identification and biodegradative characteristics of isolated microflora. *Soc. Appl. Microbiol. Lett. Appl. Microbiol.* **2007**, *45*, 87–94. [CrossRef]
32. Omar, A.M.; Taha, A.S.; Mohamed, A.A.; El-Wekeel, F. Fumigation is the ideal method in treating damaged archaeological paper using Ceratophyllum demersum L extract: A case study. *J. Basic Environ. Sci.* **2019**, *6*, 17–22.

heritage

MDPI

Article

"Like Wringing Water from a Stone!" Information Extraction from Two Rock Graffiti in North Kharga, Egypt

Nikolaos Lazaridis

Department of History, California State University Sacramento, Sacramento, CA 95819, USA; lazaridi@csus.edu

Abstract: In the course of the last ten years, the North Kharga Oasis–Darb Ain Amur Survey team, led by Salima Ikram (American University in Cairo), has been exploring a network of interconnected desert paths in Egypt's Western Desert, known as *Darb Ain Amur*. These marked paths run between Kharga Oasis and Dakhla Oasis, linking them to *Darb el-Arbain*, a notorious caravan route facilitating contacts between Egypt and sub-Saharan Africa since prehistoric times. Ancient travelers using the *Darb Ain Amur* spent several days in the midst of the Western Desert and were thus forced to use areas around sandstone rock outcrops as makeshift stopovers or camping sites. During these much-needed breaks, ancient travelers identified accessible, inscribable surfaces on the towering sandstone massifs and left on them their personalized markings. In this essay, I examine two short rock graffiti carved by such travelers in a site north of Kharga Oasis, focusing on the types of information one may extract from such ancient epigraphic materials.

Keywords: graffiti; epigraphy; desert travel; Kharga Oasis; ancient Egypt

Citation: Lazaridis, N. "Like Wringing Water from a Stone!" Information Extraction from Two Rock Graffiti in North Kharga, Egypt. *Heritage* **2021**, *4*, 2253–2260. https://doi.org/10.3390/heritage4030127

Academic Editor: Nicola Masini

Received: 23 August 2021
Accepted: 5 September 2021
Published: 7 September 2021

1. Introduction

Daring to cross the sea of sand that filled the space between Kharga Oasis and Dakhla Oasis in Egypt's Western Desert, ancient travelers had to stay on desert routes that were marked by a number of visible testimonies to earlier travelers' usage. One of these testimonies was graffiti carved on sandstone massifs' smooth surfaces. In this essay, I use a case study out of the rich corpus of ancient Egyptian rock graffiti carved by travelers who crossed these parts of the Western Desert as an opportunity to discuss the different types of information one may extract from such epigraphic materials. The selected two ancient rock graffiti come from the site of a sandstone massif that the North Kharga Oasis–Darb Ain Amur Survey team (henceforth NKODAAS) has christened "Hula Rock". NKODAAS has been exploring this area since 2005, documenting and recording different monumental and non-monumental sites on the Darb Ain Amur, a network of paths connecting Kharga Oasis to Dakhla Oasis and beyond [1,2].

Hula Rock lies approximately 34 km northwest of Kharga town and its sandstone massif runs 365 m long (Figure 1). It is situated on the Darb Ain Amur, a network of marked desert paths that begin at Ain Lebekha and end at Ain Amur, sites in which temples were built during the fourth century CE and were abandoned about a century afterwards [3]. The site was first discovered by NKODAAS in 2007 and was revisited on 22 December 2012. During the second visit, the team was able to photograph and hand-copy the two rock graffiti discussed here.

Hula Rock includes several accessible wide surfaces inviting graffiti carving on both its northwest and east faces. In addition, on the east face there is a large break in the rock, which could serve as a shelter and thus, perhaps attracted travelers' attention (Figure 2). Very close to the "shelter", NKODAAS has discovered two sets of carved one-meter-long cascading lines, which are somewhat visible at the center of Figure 2. These enigmatic carvings, which look like hula skirts (hence the name *Hula Rock*), are unique and cannot be linked with any certainty to a specific function or message [4].

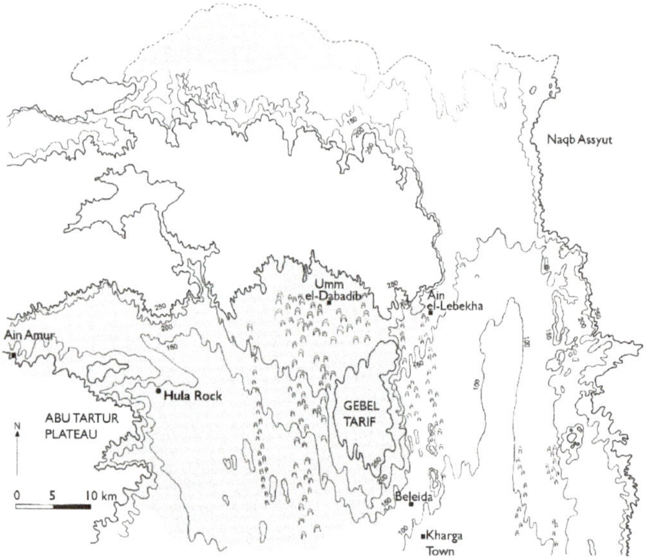

Figure 1. Map showing the location of Hula Rock (map courtesy of Nicholas Warner/NKODAAS).

Figure 2. A view of Hula Rock's east side showing the "shelter" and the figural graffito with the cascading lines (image courtesy of NKODAAS).

The site's scattered surface ceramic evidence has been dated to the Roman era (specifically to the fourth and fifth centuries CE), while there are also a few traces of New Kingdom shards [4]. The presence of the latter supports the assumed New Kingdom dates for both rock graffiti presented here, which have been based on paleographic parallels with papyrological hieratic sources (see notes 7 and 8 below).

It is, finally, worth noting that Hula Rock was visited on 1 June 1916 by A. Keatc and E. Graham, a visit that was recorded by two graffiti on the rock's east face. This visit took place two months before one or more Australian mounted units (probably belonging to the Egyptian Expeditionary Force) passed by Ain Amur. Given the troublesome times of World War I, it is doubtful that these two men were simply having a pleasant stroll in these desolate parts of the Western Desert. Instead, like their camel-riding colleagues, they too were probably members of military units using these alternative desert highways to travel northwards, where the British-led Egypt Expeditionary Force would be engaged in significant battles later that year [5] (pp. 15–32).

2. The Rock Graffiti

Hula Rock 1 (Figures 3 and 4) was carved on the rock's northwest face, 1.81 m above current ground level. It consists of a single horizontal line in cursive hieroglyphs [6], which measures 23 cm in length and 17 in maximum height. Hula Rock 1 reads:

rwḏw ḥrj
Agent Hori.

Figure 3. Hula Rock 1 (image courtesy of NKODAAS).

Figure 4. Hula Rock 1's traced copy produced by NKODAAS.

Hori's title was spelled with signs T12 and D40 [7] (p. 43). Sign T12 () included a remarkably elongated string, resembling its hieratic papyrological version attested in Papyrus Ebers, thus perhaps suggesting a Dynasty 18 date for the graffito (i.e., 1550–1292 BCE) [8] (volume 2, p. 7, sign no. 83). Hori's name was spelled with signs G5, M17, and A1. The upper part of sign A1 () is not visible, because the rock surface there is broken.

Hula Rock 2 (Figures 5 and 6) was carved on the rock's northwest face, 170 cm east of Hula Rock 1 and 1.92 m above current ground level. It consists of a single horizontal line in hieratic and measures 58 cm in length and 17 in maximum height. Hula Rock 2 reads:

zẖꜣw ꜥn-stẖ
Scribe An-Seth

Figure 5. Hula Rock 2 (image courtesy of NKODAAS).

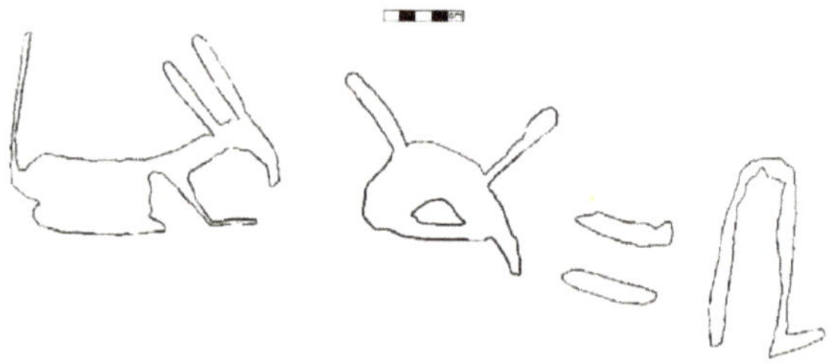

Figure 6. Hula Rock 2′s traced copy produced by NKODAAS.

An-Seth's title was spelled with the hieratic version of sign Y4. His name was spelled with the hieratic versions of signs D36, N35, D6, and E21. Note that the hieratic version of sign D6 (), which commonly replaced the hieroglyphic spelling of *ꜥn* with signs D7 or D8, included an X sign inside the eye. The vertical line over this sign's eyebrow resembles hieratic papyrological versions dated to Dynasty 18 or Dynasty 22 [8] (volume 2, p. 7, sign no. 83 and volume 3, p. 7, sign no. 83).

As somewhat visible in Figure 5, Hula Rock 2 is preceded by a 29 cm long carving of the Seth animal (sign E21:), which, based on its carving style and depth, does not seem to have been produced by An-Seth himself.

3. Discussion

At first glance, the two pharaonic rock graffiti examined here might seem too brief to contribute to NKODAAS's investigation of the desert routes in North Kharga: they, essentially, record the fleeting presence of two male Egyptian travelers of notable social status, with no reference to particular travel circumstances or to the graffiti's *Sitz im Leben*. Thus, what information can one possibly deduce from such laconic travel records?

NKODAAS's approach to interpreting ancient Egyptian rock graffiti revolves around the idea that the graffiti were carved at a specific time and on a specific rock surface in order to communicate one or more messages that given their unrestrained exposure to the eyes of passers-by, transcended the sociocultural context of meaning in which their carver operated, generating multiple interpretations. In other words, although an epigrapher's undying wish is to lay hands on the originally intended message of such a communication effort, NKODAAS's publication of such epigraphic materials attempts to offer various interpretive options whose variability depends on the ancient audience's potential reception of them. In the case of the two Hula Rock graffiti, this means that they could have potentially been considered as texts, as pictorial symbols, and/or as travel markers.

As texts, these rock graffiti publicized the two carvers' position/occupation and personal names. Hori's title is conventionally translated as "agent" or "emissary". Its translation is debatable and the attestations of this term point toward a generic meaning, used for positions in a variety of administrative contexts, such as the royal palace or the royal harem [7] (p. 43), [9] (p. 464), [10] (p. 114). In the case of Egypt's Western Desert, this title might have had specific connotations: apart from being a standard member of pharaonic expeditionary teams, such an agent was probably responsible for supervising frontier lands and maintaining the borders or managing temple estates that were far from the temples themselves [11] (p. 337), [12] (p. 107). If one contrasts the example from Parrenefer's New Kingdom stela from Abydos, in which his comparable title read *rwḏw m pr wsjr m wḥꜣt rsyt* "agent of the domain of Osiris in the Southern Oasis" [13] (p. 4), one may notice how Hori's reference is much vaguer and more succinct.

Why did Hori, following the example of many other pharaonic carvers who also omitted details such as their positions' institutional or geographical affiliations, not specify his occupational affiliations? One can only speculate that such details were not required in informal texts such as these rock graffiti, that their carvers did not have the luxury of time and space to carve longer texts that included more information, or that their carvers assumed their intended immediate audiences would have been able to guess or would have not cared about such information. Hori's decision to imitate the example of previous travelers who carved textual messages on these desert routes was an important testimony to his wish to be part of desert travelers' micro-culture, a membership that to some extent facilitated the communication of his graffito's message.

In any case, Hori's title was usually connected to an institution and thus, his visit was probably part of an official mission, a piece of historical information that suggests local or national institutions, such as the temple of Seth in nearby Dakhla Oasis or State authorities, were active in this region of the Western Desert. If NKODAAS's assumptions that Hori probably visited Hula Rock during New Kingdom's Dynasty 18 are correct, then one may correlate his presence with other evidence from Kharga Oasis attesting to especially Theban interest in this area [14], [12] (p. 107). This proves how crucial the dating of such rock graffiti is for historically contextualizing and understanding their messages, as well as for allowing NKODAAS to identify the historical usage of Darb Ain Amur's network of desert routes [15].

Like Hori, An-Seth, too, omitted any details about his occupation as a scribe. His title was extremely common in desert graffiti, since scribes were usually members of traveling teams and as teams' literati, were the best candidates for writing out (and reading out for everyone) [16] (p. 7) such textual messages. In fact, our team, along with other scholars who have examined Egyptian scribes' status, have toyed with the idea that this was perhaps not an actual title, referring to a specific official position, but an informal indication of the carver's literacy, which in the context of a society with low literacy rates, probably marked a high status [17,18].

In addition to occupational titles, both pharaonic travelers, as was the case with approximately 90% of the textual rock graffiti from North Kharga, publicized their personal names: in the case of the Hula Rock graffiti, "Hori", which meant "s/he who belongs to Horus", and "An-Seth", which can be translated "Seth is beautiful". Hori's name has been attested for both women and men in sources dating from the Middle Kingdom (i.e., roughly 2030–1650 BCE) onwards, being especially common during the New Kingdom (i.e., 1550–1077 BCE) [19] (p. 251.8). An-Seth's name, on the other hand, has not been attested so far elsewhere. There were, however, similar New Kingdom and Late Period compound names referring to female deities, such as Bastet and Mut [19] (p. 61,11 and p. 61.8), but none mentioning male deities, such as Seth.

Theophoric names such as Hori and An-Seth's could have hinted at a family's religious orientations or geographical affiliations, although one cannot be certain about such direct correlations [20]. After all, An-Seth decided to associate his textual graffito with a pictorial version of the Seth animal, a further hint at his effort to invoke his deity of preference, who was a well-known protector of Egypt's oases and desert areas [21] (pp. 105–106).

As images, such rock graffiti could also appeal to audiences who were illiterate or semi-literate in the hieroglyphic and hieratic scripts. Such travelers could have recognized specific signs, especially those whose cursive version was not very far from the hieroglyphic rendering—e.g., the falcon sign (G5) in Hula Rock 1 or the eye sign (D6) in Hula Rock 2. Such signs were sometimes targeted by image-focused visitors, either positively, as they attempted to replicate them in their own hands or negatively, as they tried to destroy them. In fact, NKODAAS has discovered at the site of Amun Rock a conspicuous example of the latter type of treatment: in the case of the rock graffito Amun Rock 9, subsequent visitors tried to destroy its eye hieroglyph [22].

Interestingly, as noted above, An-Seth's graffito was carved right next to a figural graffito depicting the hybrid animal that represented the god Seth, although there were plenty of other free spots on that rock surface to choose from. One cannot be sure which of the graffiti were carved first at Hula Rock. If An-Seth carved Hula Rock 2 after that figural graffito, then one may presume that An-Seth deliberately tried to notionally connect his message to Seth's invocation, suggesting that he, too, wanted to contribute to the figural graffito's effort to communicate with that desert deity, or that An-Seth tried to make his graffito more visible or attractive by positioning it next to a conspicuous and perhaps appealing picture. After all, Hula Rock is located relatively close to Seth Rock, a site that included several invocations of the god Seth, possibly serving as his vernacular shrine [21]. Alternatively, if the figural graffito was carved after Hula Rock 2, then one could assume that the figural graffito's carver wanted to make his graffito more visible and attractive, associating it with a well-executed text that incidentally included the picture of a Seth animal [23].

Finally, ancient or more recent travelers, regardless of their levels of literacy in the Egyptian language or of their understanding of such rock graffiti's textual meaning or pictorial symbolism, could have interpreted them as sheer markers of other travelers' fleeting presence. As such, these rock graffiti assured those who spotted them on the massifs that they were on a path that has been trodden earlier. They functioned, in other words, as desert *alamat* "signposts" [24], helping future travelers navigate these parts of Egypt's Western Desert.

The manners in which such rock graffiti have probably been received and preserved in the course of time are also delicately connected to issues of political power or cultural heritage. In the context of Darb Ain Amur's network of paths that have been used in the past by both Egyptian and non-Egyptian travelers, such graffiti indirectly propagated and confirmed Egyptian State's presence and control over these desert regions, as much as Roman authorities advertised their local presence through the establishment of forts and other monumental sites [25]. Later audiences possibly have also perceived such ancient graffiti as artifacts of ancient Egyptian heritage and thus, their treatment depended on these audiences' views of Egypt's earlier history and culture. In that respect, it is very possible that Mr. Graham and Mr. Keatc added their own graffiti on Hula Rock, at least 3100 years after Hori's visit, because they were attracted by the very presence of ancient Egyptian graffiti there and wished to associate themselves with an ancient heritage they knew about and appreciated.

4. Conclusions

Overall, communicating as texts, images, or signposts, rock graffiti like Hula Rock 1 and 2 added their voices to the muted interactions of all desert travelers who chose to carve messages on such rocks. In this way, they much resemble modern graffiti that, as defined by the art historian Franco Speroni, can be considered as expressions of connected individualities [26]. By imitating earlier desert travelers' textual and pictorial conventions, Hori and An-Seth participated in a "public forum" in which one traveler's personal identity and status was compared to, and associated with, another member of this travelers' micro-culture. The sustained flow of information between carved messages probably created a safety net that eased desert travelers' discomfort by making them feel as important contributors to the ongoing efforts to make Western Desert's terrain more hospitable and more navigable.

The NKODAAS team continues its work toward publishing the rock graffiti discovered north of Kharga Oasis, expanding, thus, the corpus of recorded ancient travelers' graffiti. Among other things, NKODAAS's forthcoming publication will include discussions about rock graffiti's relationships with informal and formal inscriptions from Egypt and about the manners in which the ancient carvers of these graffiti constructed their public identities.

Funding: This research has received funding from the National Endowment for Humanities' 2014 Scholarly Editions and Translations award, Sacramento State's Research and Creative Activity awards, the American Research Center in Egypt's Antiquities Endowment Fund, the Boyer Family, and Far Horizons.

Acknowledgments: I dedicate this short article to Ioannis Liritzis who has contributed significantly to the spread of Egyptological interest in Greece. NKODAAS is grateful to the Ministry of Tourism and Antiquities and the Kharga Inspectorate for the cooperation, help, and support.

Conflicts of Interest: The author declares no conflict of interest.

References

1. Ikram, S. The North Kharga Oasis Darb Ain Amur Survey: Past research and future questions. In *The Oasis Paper 9: A Tribute to Anthony J. Mills*; Bowen, G.B., Hope, C.A., Eds.; Oxbow Books: Oxford, UK, 2019; pp. 1–8.
2. Ikram, S. The North Kharga Oasis Darb Ain Amur Survey: Surveying the tracks between the two oases. In *The Great Oasis of Egypt: The Kharga and Dakhla Oases in Antiquity*; Bagnall, R.S., Tallet, G., Eds.; Cambridge University Press: Cambridge, UK, 2019; pp. 135–151.
3. Rossi, C.; Ikram, S. North Kharga Oasis Survey 2007–Preliminary report: Ain Lebekha and Ain Amur. *MDAIK* **2010**, *66*, 237.
4. Ikram, S. Interpreting enigmatic images: The curious case of the cascading lines. In *50 Years of Sahara: Essays in Honor of Barbara Barich*; Lucarini, G., Hamdan, M., Turjman, M., Eds.; SOR New Series: Rome, Italy, Forthcoming.
5. Woodward, D.R. *Hell in Holy Land: World War I in the Middle East*; Kentucky U.P.: Lexington, KY, USA, 2006.
6. Lazaridis, N. Carving the desert: Ancient travellers' epigraphic uses of Kharga Oasis rock surfaces. In *En détail–Philologie und Archäologie im Diskurs. Festschrift für Hans-W. Fischer-Elfert, Band 1*; Brose, M., Dils, P., Naether, F., Popko, L., Raue, D., Eds.; Zeitschrift für Ägyptische Sprache und Altertumskunde–Beihefte 7; DeGruyter: Berlin, Germany, 2019; pp. 587–600.
7. Erman, A.; Grapow, H. (Eds.) *Wörterbuch der ägyptischen Sprache, Volume 1*; Akademie Verlag: Berlin, Germany, 1971.
8. Möller, G. *Hieratische Paläographie: Die aegyptische Buchschrift in ihrer Entwicklung von der Fünften Dynastie bis zur Römischen Kaiserzeit*; Otto Zeller: Osnabrück, Germany, 1965.
9. Hannig, R. *Die Sprache der Pharaonen: Großes Handwörterbuch Ägyptisch-Deutsch*; Kulturgeschichte der antiken Welt 64; Philipp von Zabern: Mainz, Germany, 2001.
10. Quirke, S. The regular titles of the late Middle Kingdom. *RdE* **1986**, *37*, 107–130. [CrossRef]
11. Brown Wilding, M. 'Keeping enemies closer.' Ascribed Material Agency in Ancient Egyptian Rock Inscriptions and the Projection of Presence and Power in Liminal Regions. Ph.D. Thesis, Yale University, New Haven, CT, USA, 2015.
12. Long, R. Administrative control of Egypt's western oases during the New Kingdom: A tale of two cities. In *Egyptology in Australia and New Zealand 2009: Proceedings of the Conference Held in Melbourne, September 4th–6th*; Knoblauch, C.M., Gill, J.C., Eds.; BAR International Series 2355; British Archaeological Reports: Oxford, UK, 2012; pp. 105–113.
13. Darnell, J.C.; Klotz, D.; Manassa, C. Gods on the road: The pantheon of Thebes at Qasr el-Ghueita. In *Documents et Théologies Thébaines Tardives II*; Thiers, C., Ed.; *Cahiers «Égypte Nilotique et Méditeranéenne» 8*; Égypte Nilotique et Méditeranneenne and Archéologie des Sociétés Méditerranéenne: Montpellier, France, 2013; pp. 1–31.
14. Rossi, C.; Ikram, S. New Kingdom activities in the Kharga Oasis: The scribe Userhat travels westwards. *JEA* **2014**, *100*, 11–12. [CrossRef]
15. Rossi, C.; Ikram, S. Evidence of desert routes across northern Kharga (Egypt's Western Desert). In *Desert Road Archaeology*; Kuper, R., Ed.; Africa Praehistorica 27; Institut für Ur- und Frühgeschichte der Universität zu Köln: Köln, Germany, 2013; pp. 265–282.
16. McDonald, M. Ancient Arabia and the written word. In *The Development of Arabic as a Written Language*; Macdonald, M.C.A., Ed.; Supplement to the Proceedings of the Seminar for Arabian Studies 40; Archaeopress: Oxford, UK, 2010; pp. 5–28.
17. Pinarello, M.S. *An Archaeological Discussion of Writing Practice. Deconstruction of the Ancient Egyptian Scribe*; GHP Egyptology 23; Golden House Press: London, UK, 2016.
18. Ragazzoli, C. Weak hands and soft mouths: Elements of a scribal identity in the New Kingdom. *ZÄS* **2010**, *137*, 157–170. [CrossRef]
19. Ranke, H. *Die Ägyptischen Personnennamen, Band 1*; J.J. Augustin: Glückstadt, Germany, 1935.

20. Zecchi, M. Theophoric and basilophoric personal names in the Fayum in the Middle and New Kingdoms. In *Le Fayoum. Archéologie–Histoire–Religion: Actes du Sixième Colloque International, Montpellier, 26–28 Octobre 2016*; Chaufray, M.-P., Guermeur, I., Lippert, S., Rondot, V., Eds.; Harrassowitz: Wiesbaden, Germany, 2018; pp. 17–43.
21. Ikram, S. Rock art and the transformation of landscape in Kharga Oasis. In *Epigraphy through Five Millennia: Texts and Images in Context*; Dirksen, S.C., Krastel, L.S., Eds.; O. Harrassowitz: Wiesbaden, Germany, 2020; pp. 105–112.
22. Lazaridis, N. Crossing the Egyptian desert: Epigraphic work at Kharga Oasis. *Maarav J. Northwest Semit. Lang. Lit.* **2012**, *19*, 117–129.
23. Lazaridis, N. Rock communications: The interaction of textual and figural graffiti in North Kharga. In *Stone Canvas: Towards a Better Integration of "Rock Art" and "Graffiti" Studies in Egypt and Sudan*; Palkowski, P., Ed.; Institut Français D'archéologie Orientale: Cairo, Egypt, Forthcoming.
24. Riemer, H. Lessons in landscape learning: The dawn of long-distance travel and navigation in Egypt's Western Desert from prehistoric to Old Kingdom times. In *Desert Road Archaeology*; Kuper, R., Ed.; Africa Praehistorica 27; Institut für Ur- und Frühgeschichte der Universität zu Köln: Köln, Germany, 2013; pp. 77–106.
25. Rossi, C. Controlling the borders of the empire: The distribution of Late Roman "forts" in the Kharga Oasis. In *The Oasis Papers 6: Proceedings of the Sixth International Conference of the Dakhla Oasis Project*; Bagnall, R., Davoli, P., Hope, C.A., Eds.; Dakhla Oasis Project Monographs 15; Oxbow Books: Oxford, UK, 2013; pp. 331–336.
26. Ragazzoli, C.; Harmaşah, Ö.; Salvador, C. Introduction. In *Scribbling through History: Graffiti, Places and People from Antiquity to Modernity*; Ragazzoli, C., Harmaşah, Ö., Salvador, C., Frood, E., Eds.; Bloomsbury: London, UK, 2018; pp. 8–9.

heritage

Article

Archaeoastronomy of the Temples of the Bekaa Valley

Giulio Magli

Department of Mathematics, Politecnico di Milano, 20133 Milan, Italy; Giulio.Magli@polimi.it

Abstract: The Bekaa Valley, Lebanon, is famous worldwide due to the magnificent temple of Heliopolitan Jupiter at Baalbek. In recent years, new research revived the interest in the unsolved problems posed by the Baalbek monuments, including original dating and construction phases, relationships with the landscape, and nature of the cult practiced. In a preliminary paper, we used archaeoastronomy to propose that the project of the Temple of Jupiter was a unified one conceived under Herod the Great, and that the cult was strongly connected to the renewal of the seasonal cycles. Here, we extend and confirm this analysis considering the other temples of the Baalbek proper and the three prominent sanctuaries which lie in the Bekaa Valley on the way to Baalbek from Berytus, showing the existence of an orientation custom which appears to originate in Baalbek and to inform all these sacred places.

Keywords: archaeoastronomy; Temple of Jupiter Heliopolitanus; Roman temples of Lebanon

Citation: Magli, G. Archaeoastronomy of the Temples of the Bekaa Valley. *Heritage* **2021**, *4*, 1526–1537. https://doi.org/10.3390/heritage4030084

Academic Editors: Christofilis Maggidis, Nikolaos Lazaridis and Omar Abdel-Kareem

Received: 2 July 2021
Accepted: 27 July 2021
Published: 29 July 2021

1. Introduction

The Bekaa valley (Figure 1) develops for some 120 Km between Mount Lebanon to the west and the Anti-Lebanon chain to the east. At the end of the Seleucid Kingdom, the area fell into the hands of a local dynasty, the Iturean tetrarchs of Chalkis. The Roman influence in the area started with Pompey, but the tetrarchs remained in charge up to the end of the first century AD, when the foundation of a Roman colony in Berytus, the future Beirut, occurred (14 BC). It is difficult to ascertain with accuracy the exact boundaries of the colony, but there is no doubt that the valley was since then under Roman control, either directly or through the local colonial administration [1].

Between the end of the first century BC and the mid-second century AD, the valley became, for unknown reasons, a very important sacred place where huge temples and sanctuaries were constructed. Among them, by far the most important is the Temple of Jupiter Heliopolitanus, located in Baalbek at a very ideal centre of the valley. This temple is the is the core of a complex which included at least four other temples: the so-called Temples of Bacchus, of the Muses, of Venus, and a (today destroyed) Temple of Mercury. Other important, connected monuments we consider in this work are the sanctuaries/temples located in the valley along the approach to Baalbek: Niha, Hosn Niha, and Qsarnaba. This is the first systematic study on the archaeoastronomy of the Roman Temples in the Bekaa valley, while Roman urban settlements [2,3] and Nabatean tombs and settlements [4] have been studied.

All the temples of the Bekaa Valley pose problems regarding dating, attribution, and interpretation, and even the true nature of the cult of Jupiter is still debated (for a recent overview of the Archaeology of the Bekaa Valley, see [1]). The aim of the present paper is to apply the approach of modern archaeoastronomy to investigate these and related issues. In this approach, orientations are studied in context, with the objective of gaining information on the cult practiced and the chronology of the temples. Our main results are:

(1) The cult of Jupiter Heliopolitanus had an "agrarian" character related to renewal and harvesting, contrary to the sometimes claimed "solar" connotations, which instead turn out to be a possible association with the cult practiced in the so-called Temple of the Muses.

(2) The cult spread in the valley in the first half of the second century. The attribution of the large temples constructed in pre-existing sanctuaries (which is as yet unsure) is shown to be tightly related to that of the main temple as all these buildings belong to a common orientation family.

Figure 1. Positioning of the Bekaa Valley with respect to the Lebanese capital Beirut (upper left) and the main sites mentioned in the text (Image courtesy of Google Earth, editing by the author).

2. The Temples of the Bekaa Valley

In the present section we give a brief description of the monuments studied in this paper, namely the Temple of Jupiter and the so-called Temples of Bacchus, of the Muses, and of Venus in Baalbek (Figure 2) and the sanctuaries of Niha, Hosn Niha, and Qsarnaba in the Bekaa valley.

2.1. *The Temple of Jupiter at Baalbek*

The Temple of Jupiter Heliopolitanus was the largest constructed in Roman times, with a podium comparable in dimensions only with Hadrian's Temple of Venus and Rome in the Roman forum, which however is a double temple [5]. The complex develops along a monumental axis comprising a hexagonal court and Propilea, both built in the second century AD. The temple proper, which is the most ancient part, measures 48 by 90 m. It stood up to 48 m high, and its columns, resting directly on the edges of the basement, were 20 m high. The basement of the temple is sometimes called—adopting the terminology of Kropp and Lohmann [6]—Podium I, and the structure we describe here is consequently called Podium II. Contrary to my previous works on the subject [7], I came to the conclusion that this terminology is misleading and should not be used. Therefore, here the temple basement is appropriately referred to as a podium, while the other structure is likewise called an external wall (Figure 3).

Figure 2. The temples of Baalbek (Image courtesy of Google Earth, editing by the author).

Figure 3. The southeast side of the Temple of Jupiter. In the foreground, the huge blocks of the external wall, almost identical in shape and size to those of the Western tunnel in Jerusalem. The podium, with squared masonry, is also very similar to Herodian walls on the Temple Mount (photograph by the author).

The external wall runs at a distance of a few meters from the podium and is strictly parallel to it, surrounding the three sides and thus forming a giant U-shaped structure. This wall was built using the superposition of increasingly large stones as the height increases. These stones are used at the basis, larger stones are present in the second course, and huge megaliths (about 500 tons each) were used to build the third course. Finally, immense blocks—around 4 × 4 × 20 m, not less than 800 tons each—were to be placed in the uppermost course; only the southwest side was however completed, putting in place the three famous stones which are usually called (somewhat inappropriately) the "trilithon". At least three other enormous blocks remain in the quarry some hundreds of meters to the southwest [8].

Dating of the temple is very controversial, and, in view of the relevance of the monument in the history of Roman architecture, quite important. The unique aspect which appears to be relatively sure is that the erection of the columns was almost completed in the year 60 AD, as a graffito left by stonemasons has been found on one of the column's summits suggesting this date [1]. Thus, the final phase of construction—with the erection of the enormous columns' drums—belongs to the Julio-Claudian period. Consequently, the most accepted archaeological viewpoint was that the whole building is Julio-Claudian. However, the pre-existence of a Hellenistic sanctuary would seem likely, and a recent architectural analysis of the podium has shown striking similarities to Herodian sanctuaries, such as the use of alternating rows of headers and stretchers and the presence of drafted-margin masonry. In particular, obvious similarities to the Herodian walls at the Temple Mount at Jerusalem exist, not only in general appearance, but also in proportions and measures. All in all, it is natural to conclude that the podium was originally built by Herodian architects [6]. According to the same authors, the external wall postdates the Herodian phase and was designed to "harmonise" the dimensions to Roman standards. However, as I have recently pointed out [7], there also exist clamorous analogies with Herodian buildings for the external wall, in particular between the intermediate course of megaliths and the one located in the western tunnel in Jerusalem; furthermore, the "style" of the immense trilithon's blocks also demonstrates a clear parallel in the Temple Mount walls, for instance in Barclay's gate: In both cases the size of the blocks is similar, and the method of shaping the blocks is also similar. Admitting that the external wall and the podium were built together would also explain their otherwise inexplicable lack of structural connection (the alleged Roman enlargement would have been much easier to obtain by adding masonry to the existing basement) with the intention of building a covered-gallery ambulacrum in the original project. This might be related to oracular activities which were held at the temple, as recorded by Macrobius (Saturnalia I.23.14–16) who states that the Jupiter of Heliopolis announced a bad omen about Trajan's expedition in the Parthian Empire. Finally, it would also explain the identical orientation of the two structures (see Section 4).

As far as the proposal of Herod's intervention is concerned, it should be recalled that Baalbek was under Roman rule and not enclosed in Herod's reign. Since the dismantling of the Iturean control of the area by the Romans, however, we know from several reports that Herod the Great, the vassal King of neighbouring Judea, showed keen attention to the Roman possessions, contributing to relevant architectural projects in various towns including Berytus and Samaria (renamed by him to Sebaste in honour of Augustus).

Other interesting problems occur when we come to analyse the god worshipped in the temple, a question of course related to the chronology as well. The god was called Jupiter Optimus Maximus Heliopolitanus (his name is attested in many sources) and was therefore a form of the Roman Jupiter. However, his iconography—known from cult statues and reliefs and described in the unique written texts we have about Baalbek, with Porphyry's description reported by Macrobius in the fourth century AD—is very far from that of "standard" Jupiter [9,10]. The Heliopolitan God is young and beardless, with a vase-shaped hair volume (Figure 4). Decorations of it vary, but most common are grain ears, although eagles, sun discs, and others also occur. The attributes are a whip in the raised right hand

and again grain ears in the lowered left hand. The god is accompanied by two walking bulls facing the same direction and always wears a sort of tight-fitting robe which exabits complex, and multi-variated, decorations, although the front side usually has Helios and Selene in the top register, and a lion head or lion head mask is often depicted at the bottom.

The origin of this curious iconography has been sometimes attributed to tribal, pre-Roman deities, but there is no known pre-Roman image of this kind, although "standard" Zeus is documented in Iturean coins [9]. Another traditional interpretation is in terms of a solar deity, but besides the mysterious toponym "Heliopolis" attributed to Baalbek (the original Heliopolis of course was the main centre of the cult of the sun god Ra in ancient Egypt), we do not have any proof whatsoever of such a solar character. To complicate matters, research has been delayed for years by the idea that a triad of gods—Jupiter, Venus and Mercury—was actually worshipped in Baalbek trough a syncretism with local deities (Jupiter with Hadad, an ancient Mesopotamian God of Thunder, and Venus with Atargatis, a marine deity). Only recently has this idea of a "Heliopolitan Triad" been shown to lack solid evidence [8].

Figure 4. A bronze statuette of Jupiter Heliopolitanus (image in the public domain/Wikimedia Commons).

2.2. The Other Baalbek Temples

Parallel and extremely close to the Temple of Jupiter stands the so-called Temple of Bacchus, another masterpiece of Roman architecture. It is slightly larger than the Parthenon, with a similar external colonnade. The high staircase with an adyton appears instead to be of local character and is replicated in other sanctuaries. Its attribution to Bacchus is based on the interpretation of some of its reliefs and is purely hypothetical. Other decorations seem to refer to Mercury (but this would give two temples to this deity in Baalbek) yet others to Venus, not to mention that it may be a second temple dedicated to Jupiter Heliopolitanus. Its date of construction should be around the first half/middle of the second century [1].

Close to the entrance of the Jupiter complex, a terraced area contains two further temples [11,12]. The most ancient of the two, today in ruins, is the so-called Temple of the Muses. The suggested date for this building is the end of the first century BC which would make it the very first known Baalbek temple. The Temple was pseudoperipteral, with Corinthian columns inserted in the walls of the cella, following a style present in Italy and Gaul. Curiously, the temple was built in a depression and was—at least according to its excavators—regularly flooded up to the summit of the podium, at least until the construction of an enclosure retaining wall occurred in the second half of the second century. The axis of the Temple is not parallel to that of the later Temple of Jupiter (see next section). Attribution of this temple is indirect and very doubtful (it relies only upon an inscription which prizes a citizen for financing a drainage channel to protect the "Temple of the Muses"; however, the inscription is located on the neighbouring Temple of Venus, which—to mention only one of the problems—is also not securely attributed).

The Temple of Venus is a prostyle round Temple built on the same temenos enclosure of the much earlier Temple of the Muses. It was built on a sort of double podium, but it had problems of construction and was perhaps never finished. Attribution to Venus is very poorly grounded (some external niches are decorated with seashells). It is clearly a late building, perhaps even of the third century AD.

Finally, it is certain that another temple, probably dedicated to Mercury, was located on a hill called Sheikh Abdallah to the southeast of the Jupiter temple entrance [13]. Nothing remains of this structure, however, besides a part of a monumental stairway 13 m wide which ended in a propylon to the temple enclosure.

2.3. The Three Sanctuaries to the South of Baalbek

Some 25–30 Km southwest of Baalbek lie three monumental sanctuaries: Niha, Hosn Niha, and Qsarnaba [14,15]. They are not located directly in the valley but rather on its western side. Niha and Hosn Niha appear to have been originally connected with existing springs; all three are sumptuous buildings, the style and architecture bearing many similarities to the much more famous Baalbek Temple of Bacchus. The Niha complex comprises two temples and a sort of altar platform located close to the point in which the axes of the two Temples cross. Perennial springs are present nearby, and a river flows between the two. The platform should pre-date the temples, of which the smaller (temple B) is a tetrastyle prostyle 12 × 27.5 m, built on a levelled terrace. The so-called Great Temple or Temple A is built on a podium. It is a huge tetrastyle prostyle, 18 by 41 m, with an adyton shrine. The date of construction of both temples is unsure, but Temple B should pre-date Temple A, with the latter built in the first half of the second century AD. The iconography present in the temple (on lintels, walls and altars) is of difficult interpretation. For instance, a relief shows a man, probably a priest, flanked by two victories and holding cult objects, perhaps a branch or an ear of corn. In another relief, three figures, including a priest, attend at a libation or sacrifice. Uniquely, the style clearly recalls the iconography associated with Jupiter Heliopolitanus, showing close ties with Baalbek. As far as the specific dedication of the temples is concerned, attribution is unknown, but there exist hints that the worshipped deities were Hadaranes and Atargatis. In particular, a votive cippus and a limestone stela have been found, bringing inscriptions honouring a priestess of these two deities. Interestingly, on the cippus Hadaranes is represented with a style, which again directly recalls Jupiter's iconography.

The complex of Hosn Niha lies close to Niha to the northeast. The sanctuary is located on a terrace levelled with the help of retaining walls. As in Niha, the focus of attraction is a "great" temple (temple A), probably built in the early second century in a pre-existing sacred area which already included a sacred spring and a "small" temple. The small temple is squared, in antis, with pronaos and cella. The great temple is a tetrastyle prostyle with Corinthian columns, pronaos, and cella. The temple (dimensions about 14 × 28 mts) lies on a podium and is accessed through a large, monumental staircase. Huge stones were used for the walls of the cella. All in all, this temple also shares many similarities with the so-called Temple of Bacchus at Baalbek and thus with Niha. Some 200 m to the southwest, the remains of a further building (called the double sanctuary) can be seen. The sanctuary consists of two rooms, both with a single entrance that faces towards the main sanctuary. In front of the entrance to the northernmost room are the remains of two altars, each consisting of a large, squared stone block and an associated top block. Attribution of the Hosn Niha complex is doubtful, but there exists the possibility of a pair of deities, possible a divine couple with a deity and consort. As such this raises the possibility that this is another temple dedicated to Hadaranes and Atargatis, the deities perhaps worshipped at Niha.

The Temple of Qsarnaba is located some 3.2 km northeast of Niha. The temple, similar to the Great Temple of Niha, is built on a high platform and is reached by a high flight of steps. The structure is collapsed but originally was a hexastyle prostyle, comprising a pronaos and a cella with an underground crypt. It is difficult to ascertain if other buildings were present, due to modern constructions.

3. Orientations of the Temples of the Bekaa Valley

Archaeoastronomy has advanced since the times of the mere and somewhat naive studies or uncontextualised alignments and can be considered today as a complete, interdisciplinary science tightly connected to cognitive archaeology [16]. Most archaeologists have started to value this discipline for its cognitive character. One key example, which is relevant in our context, is that of the Greek temples. From a general—and vague—notion that "most Greek temples face the rising sun", studies have turned to analysing specific orientations to connect them to specific deities, sacred attributes, and rites [17–22]. It is our aim here to approach the problem of the archaeoastronomy in the Bekaa valley exactly in the same way, so that—for instance—vague statements such those present in many publications on Baalbek where it is noticed that the Temple of Jupiter "opens to the east" as alleged proof of the solar character of the deity can be overcome by a rigorous analysis.

To study orientations, we use here satellite imagery (Google Earth) which usually allows the determination of both the azimuth and the horizon height (here defined as viewed from the inside looking out) with good approximation. For the present paper, considering that the satellite mapping of the area out of Baalbek proper is not at the maximal possible definition, it is safe to assume an accuracy of ±1 degree. To study the sky at the times of construction, we used the program Starry Night Pro. A complete list of orientations is given in Table 1, where declinations are also calculated using the program GETDEC (kindly provided by Clive Ruggles) which takes into account refraction. To simplify reading of the data, a diagram of orientation is given in Figure 5.

Table 1. Orientations of the temples of the Bekaa Valley.

Place	Date	Temple	Azimuth	Horizon Height	Declination	Notes
Baalbek	Approx. 10 BC (Herodian phase) 50 AD (Julio-Claudian phase)	Jupiter	75°30′	5°	14°40′	
Baalbek	Mid-second century AD	Bacchus	75°30′	4°30′	14°20′	
Baalbek	Third century AD	Venus	320°			towards Temple of Jupiter
Baalbek	Second half of the first century BC	Muses	63°	3°40′	24°	summer solstice sunrise
Baalbek		Mercury				not measurable
Niha	Mid-second century AD	A	74°	5°	16°	
Niha	First century AD	B	156°	2°	−47°46′	
Hosh Niha	Mid-second century AD	A	76°	3°50′	13°34′	
Hosh Niha	First century AD	B	156°	7°30′	−42°40′	
Hosh Niha	First century AD	Double Sanctuary	67°30′	0°	18°10′	orthogonal to temple B; lunar geocentric dec. 18°42′
Qsarnaba	Mid-second century AD		76°30′	2°	12°30′	
Qasr el Banat			75°	1°	12°45′	

The orientation of the Temple of Jupiter already poses interesting questions. The azimuth is of 75°30′ which, taking into account the horizon height, gives a declination ~14°44′. This declination is within the solar range but of course does not correspond either to equinoxes or to the summer solstice; the sun rises in alignment with the temple in early May/mid-August (Gregorian, but the difference with the Julian in the first/second century AD was negligible). These dates are not of special significance for the solar cycle, as a confirmation of the mentioned doubts on a "solar" Jupiter. The dates are not notably close to days of special significance in the Roman calendar either.

A topographical alignment with the area at which the Hellenistic water pipe enters the city, Ain Juj, has been proposed [23]. At Ain Juj the remains of a small round building, with suggested date at the end of 1 BC, were found, but the building is today lost, and its location is unsure. Furthermore, the idea that the largest Temple of the Roman empire was oriented to a water supply, however sacred its source might have been, is frankly difficult to believe (the assertion that the so-called Temple of the Muses is also directed to the same point is only approximatively true; the orientation of this temple is instead clearly astronomical, as shown below). The possibility thus remains that the intended alignment was stellar, and indeed a quite important celestial object was rising in alignment with the temple: the Pleiades. Of course, the Pleiades is an asterism, not a single star (seven stars can be distinguished with the naked eye); however, they can be considered (and were considered in antiquity) as a single entity. They occupy a portion of the sky which spans about ~1/2° in declination. Their declination in Herodian times was between 15°30′ and 16°, slowly increasing with time. The agreement with the temple declination is therefore good, and the horizon height which corresponds to the temple front assures that the asterism was visible. Is this orientation in accordance with what we know about the Heliopolitan Jupiter?

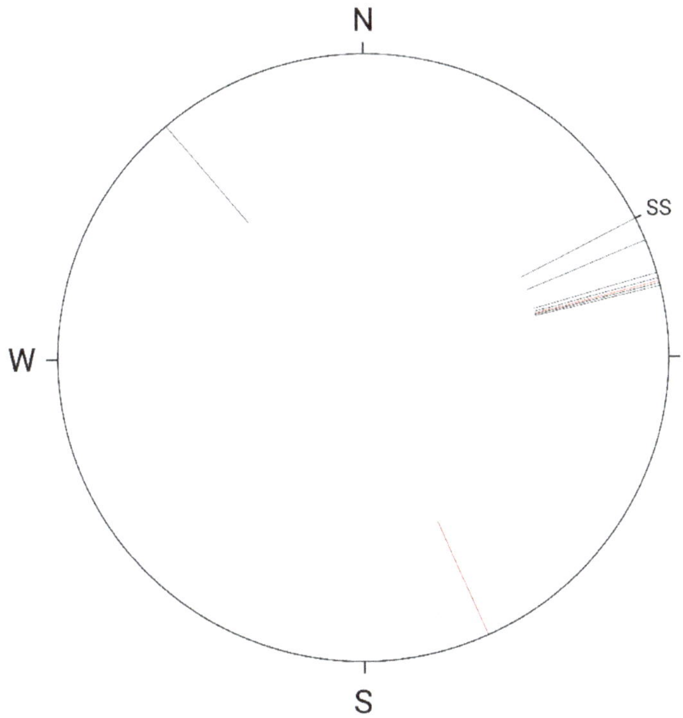

Figure 5. Diagram of orientation of the temples of the Bekaa Valley.

Interest in the Pleiades is well documented in the Greek religion; for instance, the role of this asterism has been shown to be fundamental for the rites of the Artemis Orthia sanctuary in Sparta [17]. This asterism was associated—already in the Hesiod calendar, 8th century BC—with the harvest time of the cereals, indicated by their heliacal rising in the first week of May. This is also the period in which the phenomenon occurred in Baalbek at the end of the first century BC. Therefore, the alignment of the temple individuated both the direction of the heliacal rising of the Pleiades and that of sun at rising, a few minutes later, on the same days, a quite peculiar coincidence. All in all, the alignment of the temple actually points to "agrarian" iconographical associations of Heliopolitan Jupiter, and in particular to renewal and harvesting. The "solar" character of Jupiter Heliopolitanus is therefore indirect; in some way it brings to mind some peculiarities of the cult of Mithra. Many details of Mithraic mysteries are unknown and subject to debate, but the "friendship" of the god with Helios is represented in Mithraic iconography as well as in Jupiter's, and the Mithraic sacrifice of the bull, the central scene of any mithraeum, is certainly connected to renewal and harvesting, as ears of wheat are seen coming out from the bull's tail or wound [24]. The possible existence of a cult for a "true", different solar deity in Baalbek remains due to the orientation of the so-called Temple of the Muses, the attribution of which is unclear. With a declination of ~24°, the temple is indeed clearly oriented to the rising of the Sun at summer solstice.

If we now extend the analysis to the other sanctuaries of the valley, we can notice that a very peculiar situation occurs. Indeed, the original Temples of Niha and Hosn Niha, probably devoted to the cult of waters, were oriented in an identical manner—towards the southeast and the valley. It is interesting to note that, either by chance or by design, their orthogonal direction points to the Moon rising at the minor lunar standstill, and this is also the declination of the double sanctuary at Hosn Niha. Be that as it may, when it was decided to build in both centres of the cult a new, massive temple in the style of the Temple of Bacchus in Baalbek, the old orientation was changed, and the new orientation was also shared by the third temple constructed along the same architectural lines, Qsarnaba. The Temples of Niha A, Hosn Niha A and Qsarnaba indeed all belong to a very limited range of azimuths (74° to 76°30′) and declinations (12°30′ to 16°) which includes the values for Jupiter and Bacchus as well (what appear to be the remains of a further temple in the same style located at Qasr el Banat are also measurable and give similar results). Of course, the sample of data is too small to perform a meaningful statistical analysis, but the fact that three out of three of the temples built in the Valley in the same period and with similar architectural characteristics of the Temple of Bacchus all share a very similar orientation, and that this orientation is the same of the pre-existing main sanctuary in Baalbeck, is a very clear hint (if not proof) of their Heliopolitan character.

4. Discussion and Conclusions

As in recent approaches to the orientation of the Greek temples, we have shown here that the astronomical symbolism incorporated in the temple architecture can give information on the god or cult to which the temple was dedicated. In particular, Heliopolitan Jupiter, with his association to renewal and harvesting, also appears to have been associated with the sun rising in May and the reappearance of the Pleiades. Both the podium and the external U-shaped wall were built respecting closely this orientation, and this is a further hint to a unitarity of the project in Herodian times, since stellar alignments change due to precession, and on the occasion of the alleged second-phase enlargement we would also expect a change in the corresponding alignment.

The comparison of this alignment with those of all other related temples leads to several interesting results. First of all, while there is no proof of a "solar" character sometimes attributed to Jupiter Heliopolitanus, a temple clearly oriented to the sun rising at the summer solstice does exist in Baalbeck: the so-called Temple of the Muses, which is also the oldest. This is a hint to a solar deity worshipped there.

Another result obtained here is the fact that the orientation of the main temple turns out to be shared by the so-called Bacchus temple and by the temples constructed in the valley in the same style. They form a family of monuments, the attribution of which is unsure, but their shared orientation is a strong hint to their tight dependence on the main cult.

All in all, and notwithstanding the mentioned difficulties and the lack of details about the evolution of the cults practiced in the valley, archaeoastronomy can be used to contribute to a tentative and incomplete "cognitive" chronology of the Roman temples of the Bekaa valley as follows:

(1) In pre-Roman or early Roman times, the so-called Temple of the Muses of Baalbek was a place of worship of a deity that very probably had solar connotations due to its orientation at summer solstice sunrise.

(2) During Herod's time, that is under early Roman rule, something happened that made Baalbek the cultic centre of quite an original "version" of Jupiter which had agrarian characteristics associated with renewal and fertility.

(3) Herod's architects started the project of a huge temple. The temple is oriented to the rising of the Pleiades in early May, a phenomenon associated with harvesting since Hesiod's times. The platform of the temple should boast a U-shaped (covered) ambulacrum which was projected in a way very similar to the walls today visible in the Western Tunnel and Barclays Gate of the Temple Mount in Jerusalem, using

immense blocks. The podium and what today appears as an external wall clearly share the same orientation.

(4) In Julio-Claudian times the columns were erected and other works were made, but the megalithic project of the ambulacrum was left unfinished.

(5) Around the mid-second century AD (probably under Antoninus Pius), the so-called Temple of Bacchus was built. The temple is very close to the Temple of Jupiter and strictly parallel to it; the mentioned difficulties in establishing a secure attribution may arise simply from the fact that it was dedicated to Jupiter as well.

(6) The cult spread in the already existing sanctuaries in the valley, which were originally devoted to the cult of waters. As a consequence of this spread, the great Temples of Niha A, Hosn Niha A and Qsarnaba were built with "Heliopolitan" characteristics similar to those of the Temple of Bacchus. In particular, the orientations of these temples adhere to the Heliopolitan model and are radically different from those of the pre-existing buildings (like the Niha Temple B) at the same places.

Funding: This research received no external funding.

Institutional Review Board Statement: Not applicable.

Acknowledgments: It is an honour for me to dedicate the present work to my friend and colleague Ioannis Liritzis on the occasion of his retirement from the University of the Aegean.

Conflicts of Interest: The author declares no conflict of interest.

References

1. Paturel, S. *Baalbek-Heliopolis, the Bekaa, and Berytus from 100 BCE to 400 CE A Landscape Transformed*; Brill: Leiden, The Netherlands, 2016; Volume 426.
2. Rodríguez-Antón, A.; González-García, A.C.; Belmonte, J.A. An archaeoastronomical approach to roman urbanism. Orientation of roman settlements across the empire. *Mediterr. Archaeol. Archaeom.* **2018**, *18*, 89–95.
3. Rodríguez-Antón, A.; Belmonte, J.A.; González-García, A.C. Romans in the near east: The orientation of roman settlements in present-day Jordan. *Mediterr. Archaeol. Archaeom.* **2016**, *16*, 153–160.
4. Liritzis, I.; Al-Otaibi, F.M.; Castro, B.; Drivaliari, A. Nabatean tombs orientation by remote sensing: Provisional results. *Mediterr. Archaeol. Archaeom.* **2015**, *15*, 289–299.
5. Segal, A. *Temples and Sanctuaries in the Roman East: Religious Architecture in Syria, Iudaea/Palaestina and Provincia Arabia*; Oxbow Books: Oxford, UK, 2013.
6. Kropp, A.J.M.; Lohmann, D. 'Master, look at the size of those stones! Look at the size of those buildings!' Analogies in Construction Techniques Between the Temples at Heliopolis (Baalbek) and Jerusalem. *Levant* **2011**, *43*, 38–50.
7. Magli, G. The Archaeoastronomy and Chronology of the Temple of Jupiter at Baalbek. In *Archaeoastronomy in the Roman Worl*; Magli, G., González-García, A.C., Belmonte Aviles, J., Antønello, E., Eds.; Springer: New York, NY, USA, 2019; pp. 245–251.
8. Abdul Massih, J. The Megalithic Quarry of Baalbek: Sector III the Megaliths of H. ajjar al-H. Ibla. *J. East. Mediterr. Archaeol. Herit. Stud.* **2015**, *3*, 313–329.
9. Kropp, A.J.M. Jupiter, Venus and Mercury of Heliopolis (Baalbek). The images of the "triad" and its alleged syncretisms. *Syria* **2010**, *87*, 229–264.
10. Kropp, A.J.M. The cults of Ituraean Heliopolis (Baalbek). *J. Rom. Archaeol.* **2009**, *22*, 365–380. [CrossRef]
11. Hoebel, F. The Sanctuary in the Venus Area—Cult Topography and Urban Development. In *Baalbek/Heliopolis: Results of Archaeological and Archi#tectural Research 2002–2005*; van Ess, M., Ed.; BAAL: Beirut, Lebanon, 2006; pp. 161–177.
12. Hoebel, F. Zwischen Orient und Okzident: Die Kultbauten im Areal Santa Barbara. In *Baalbek-Heliopolis 10000 Jahre Stadtgeschichte*; van Ess, M., Rheidt, K., Eds.; Philipp von Zabern: Darmstadt, Germany, 2014; pp. 80–91.
13. Buthcher, K. Heliopolis coins, temples, and sightlines. *Bull. Inst. Class. Stud.* **2019**, *62*, 106–115. [CrossRef]
14. Aliquot, J. *La Vie Religieuse au Liban sous l'Empire Romain, Bibliothèqu Archéologique et Historique*; Institut Français du Proche-Orient: Beirut, Lebanon, 2009.
15. Yasmine, J. The Niha Sites (Lebanon) Cultural Landscape: A 3D Model of Sanctuaries and Their Context. *Int. Arch. Photogramm.* **2013**, *XL-5*, 699–704.
16. Magli, G. *Archaeoastronomy. Introduction to the Science of Stars and Stones*; Springer: New York, NY, USA, 2015.
17. Boutsikas, E.; Ruggles, C.L.N. Temples, stars, and ritual landscapes: The potential for archaeoastronomy in ancient Greece. *Am. J. Archaeol.* **2011**, *115*, 55–68. [CrossRef]
18. Hannah, R. Greek temple orientation: The case of the older Parthenon in Athens. *Nexus Netw. J.* **2013**, *15*, 423–443. [CrossRef]
19. Castro, B.M.; Liritzis, I.; Nyquist, A. Oracular Functioning and Architecture of Five Ancient Apollo Temples Through Archaeoastronomy: Novel Approach and Interpretation. *Nexus Netw. J.* **2016**, *18*, 373–395. [CrossRef]

20. Liritzis, I.; Vassiliou, H. Archaeoastronomical Orientation of Seven Significant Ancient Hellenic Temples. *Archaeoastron. Int.* **2003**, *17*, 94–100.
21. Hannah, R. The roles of observational astronomy in ancient Greece. *Sci. Cult.* **2015**, *1*, 47–56.
22. Hannah, R.; Magli, G. A Orlando Astronomy, topography and landscape at Akragas' Valley of the Temples. *J. Cult. Herit.* **2017**, *25*, 1–9. [CrossRef]
23. Rheidt, K. *Remarks on the Urban Development of Baalbek. Baalbek/Heliopolis: Results of Archaeological and Architectural Research 2002–2005*; BAAL: Beirut, Lebanon, 2006.
24. Beck, R. *The Religion of the Mithras Cult in the Roman Empire: Mysteries of the Unconquered Sun*; Oxford UP: Oxford, UK, 2007.

MDPI
St. Alban-Anlage 66
4052 Basel
Switzerland
Tel. +41 61 683 77 34
Fax +41 61 302 89 18
www.mdpi.com

Heritage Editorial Office
E-mail: heritage@mdpi.com
www.mdpi.com/journal/heritage

CPSIA information can be obtained
at www.ICGtesting.com
Printed in the USA
LVHW071500100422
715714LV00041B/689